SpringerBriefs in Philosophy

For further volumes:
http://www.springer.com/series/10082

Jan Willem Wieland

Infinite Regress Arguments

 Springer

Jan Willem Wieland
VU University Amsterdam
Amsterdam
The Netherlands

ISSN 2211-4548   ISSN 2211-4556   (electronic)
ISBN 978-3-319-06205-1   ISBN 978-3-319-06206-8   (eBook)
DOI 10.1007/978-3-319-06206-8
Springer Cham Heidelberg New York Dordrecht London

Library of Congress Control Number: 2014936436

Printed on acid-free paper

Springer is part of Springer Science+Business Media (www.springer.com)

Contents

Chapter 1
Introduction

1.1 Overview

Infinite regress arguments (henceforth IRAs) are powerful arguments and frequently used in all domains of philosophy. They play an important role in many debates in epistemology, ethics, philosophy of mind, philosophy of language, and metaphysics. In historical debates, they have been employed, for example, against the thesis that everything has a cause and against the thesis that everything of value is desired for the sake of something else. In present-day debates, they are used to invalidate certain theories of justification, certain theories of meaning, the thesis that knowledge-how requires knowledge-that, and many other things.

Indeed, there are infinite regresses of reasons, obligations, rules, relations and disputes, and all are supposed to have their own moral. Yet most of them are involved in controversy. Hence the question is: what exactly is an IRA, and how should such arguments be evaluated? This book answers these questions and is composed as follows:

- Chapter 1 introduces the topic, presents an overview of classic IRAs, and explains what we should expect from theories about IRAs.
- Chapters 2 and 3 summarize two theories of IRAs that have been presented in the literature: namely The Paradox Theory and The Failure Theory.
- Chapters 4 and 5 concern two case studies which illustrate the general insights about IRAs presented in the previous chapters in some detail.

This book is of metaphilosophical nature. It is about how philosophers *can* and *should* proceed in their inquiries. Indeed: what steps can and should philosophers take to defend their views? IRAs, then, are arguably one of the main philosophical argumentation techniques (that is, next to thought experiments).

The book's ultimate aim is to make a difference to the philosopher's practice. Particularly, the hope is that from now on disputes about any particular IRA (concerning what it establishes, or concerning whether it can be resisted) will be more clearly motivated, and indeed more clearly framed, in terms of the guidelines outlined in what follows.

J. W. Wieland, *Infinite Regress Arguments*, SpringerBriefs in Philosophy, DOI: 10.1007/978-3-319-06206-8_1, © The Author(s) 2014

I am grateful to many colleagues for advice. Special thanks go to: Arianna Betti, Bert Leuridan, Anna-Sofia Maurin, Francesco Orilia, Benjamin Schnieder, Maarten Van Dyck, Erik Weber, and René van Woudenberg. Thanks also to Jon Sozek for the language edit, to the Flemish and Dutch science foundations for financial support, and to the referees and editors of my articles on which this book is based (see Wieland 2013a, b, c, 2014).

1.2 Two Examples

To introduce the topic, let us consider two classic examples: Juvenal's guardians, and the ancient (yet timelessly relevant) Problem of the Criterion.[1] Famously, the Roman poet Juvenal posed the question: "But who will guard the guardians?" (*Satire* 6.O29-34). The IRA suggested by this question could be spelled out in two different ways:

> Suppose you want to have your partner guarded so that he or she can no longer commit unfaithful acts. As a solution, you hire a guardian. Yet, as it happens with guardians, they cannot be trusted either. So a similar problem occurs: you want to have the guardian guarded. As a solution, you hire another guardian. Regress. Hence, hiring guardians is a bad solution to have your partner guarded.

> Suppose that your partner is unreliable, that all unreliable persons are guarded by a guardian, and that all guardians are unreliable. This yields a regress which is absurd. Hence, either it is not the case that all unreliable persons are guarded by a guardian, or it is not the case that all guardians are unreliable.

In the second case, but not in the first, the notion of 'absurdity' has an important role. The fact that a certain regress is absurd does not follow from assumptions that generate the regress, and has to be argued for independently. Moreover, if the regress is not absurd, then none of those assumptions have to be rejected. The differences between these reconstructions will become clear later on, yet, as we will see, in principle both are legitimate reconstructions. Here is the Problem of the Criterion drawn from the ancient sceptics, a well-known case that inspired many further similar problems:

> In order to decide the dispute that has arisen [...], we have need of an agreed-upon criterion by means of which we shall decide it; and in order to have an agreed-upon criterion it is necessary first to have decided the dispute about the criterion. [...] If we wish to decide about the criterion by means of a criterion we force them into infinite regress. (Sextus, *Outlines of Pyrrhonism*, 2.18–20)

Suppose, for example, that you want to settle a dispute about whether Juvenal had a wife. To do so, you introduce another proposition, such as that Juvenal has been banished his whole life and so could not have had a wife. Of course, this

---

[1] For the first example, cf. Hurwicz (2008). For the second, cf. Chisholm (1982, Chap. 5), Amico (1993), Lammenranta (2008).

second proposition is disputable too. To settle that new dispute, you introduce a third proposition, say that the sources about Juvenal's banishment are highly reliable. Of course, this third proposition is disputable too, and so on into the regress.

Generally, the setting is you have to decide whether a certain proposition p is true. You can do this critically, i.e. by a proof, or uncritically. If you do it uncritically, then your decision is arbitrary and will be discredited. But if you do it critically and use a criterion c_1 to decide whether p is true, you need first to decide whether c_1 correctly rules what is true and what is not. Again, there are two options: you can do this critically, or not. If the latter, your decision will be discredited. So you do it critically, and refer to a meta-criterion c_2 which determines what criteria correctly determine what is true and what is not. But now you need first to decide whether c_2 correctly determines the correct criteria. Again, you can do this critically, or not. And so on.[2]

Again, the pattern can be described in two, slightly different ways:

Suppose you want to settle the dispute about some issue. As a solution, you introduce a criterion on the basis of which it can be settled. Yet, that criterion is disputable too. So a similar problem occurs: you want to settle the dispute about that criterion. As a solution, you introduce another criterion. Regress.

Suppose that at least one dispute is settled, that disputes are settled only if there is an agreed-upon criterion on the basis of which they are settled, and that there are agreed-upon criteria only if the disputes about them are settled. Regress.

Now the question is what conclusion should be drawn from such a regress of criteria. Does it follow that one cannot settle *all* disputes? Or that one cannot settle even *one* dispute? The second conclusion is clearly stronger than the first: if one fails to settle all disputes, one might still settle many of them; yet if one fails to settle any dispute, one automatically fails to settle all of them. As we will see in this book, there are in fact four possible conclusions IRAs can have (and all require their own kind of support):

- **to refute an existentially quantified statement**, for example the statement that at least one dispute is settled;
- **to refute a universally quantified statement**, for example the statement that for all disputes it holds that they are settled only if there is an agreed-upon criterion to settle them;
- **to demonstrate that a certain solution fails to solve a universally quantified problem**, for example that the given solution fails to settle all disputes;

---

[2] Actually, the Problem of the Criterion might also involve a *circularity*, rather than a regress. In that case, you prove that c_1 correctly determines what is true and what is not by showing that it predicts the right results. Here, you already know what is true and what is not, and so whether p is true or not. This is circular, for we started from the situation where you still have to decide whether proposition p is true.

- **to demonstrate that a certain solution fails to solve an existentially quantified problem**, for example that the given solution fails to settle even one dispute.

Thus IRAs can be used for these four different purposes. Before explaining in Chaps. 2 and 3 how IRAs of these four kinds work, I will provide an overview of twelve main classic IRAs. In this overview, I will sidestep the issue in which of the four ways the given case should best be understood, and simply opt for one specific reconstruction (but, as will become clear later, alternative reconstructions of the cases are possible).

1.3 Overview of Classic Cases

Plato's Third Man

Suppose anything is large only if it partakes of the form Largeness, that Largeness is itself large, and that forms are distinct from anything that partakes of them. This yields a regress that is absurd. Hence: it is not the case that anything is large only if it partakes of the form Largeness.
Source: *Parmenides*, 132a–b

Aristotle's Highest Good

Suppose anything is good only if we desire it for the sake of something else that is good. This yields a regress that is absurd. Hence: at least one thing is good and not desired for the sake of something else that is good, i.e. the highest good that is desired for the sake of itself.
Source: *Nicomachen Ethics*, 1094a

Sextus' Reasons

Suppose that some propositions are justified to someone and that propositions are justified to someone only if that person has reasons for them which are themselves justified to that person. This yields a regress that is absurd. Hence: either it is not the case that all justified propositions are supported by reasons which are themselves justified propositions, or no proposition is justified to anyone.
Source: *Outlines of Pyrrhonism*, 1.166–7

Aquinas' Cosmological Argument

Suppose that every finite and contingent being has a cause, and that every cause is a finite and contingent being. This yields a regress that is absurd. Hence: it is not the case that every cause is a finite and contingent being. There must be a first cause which is not finite or contingent, namely God.
Source: *Summa Theologica*, book I, v. 2, Sect. 3

Hume's Induction

Suppose you want to justify an inductive inference. As a solution, you rely on the assumption that the future resembles the past. Yet, you cannot justify the initial inference unless you justify this assumption. As a solution, you again rely on the assumption that, in this case too, the future resembles the past. Regress. Hence: you will never justify any inductive inference.
Source: *An Enquiry Concerning Human Understanding*, Chap. 4

Bradley's Unity

Suppose you have to explain how a relation R forms a unity with its relata a and b. As a solution, you invoke an additional relation R* that is supposed to unify R with a and b. Yet, you do not explain how R forms a unity with its relata unless you explain how R* forms a unity with its relata R, a and b. As a solution, you invoke yet another relation R** to unify R* with its relata. Regress. Hence: you cannot explain how any relation forms a unity with its relata.
Source: *Appearance and Reality*, Chaps. 2–3

Carroll's Tortoise

Suppose you have to demonstrate that a conclusion follows from certain premises. As a solution, you add an extra premise to the argument: 'if the foregoing premises are true, the conclusion is true.' Yet, you do not show that the conclusion follows unless it is shown that it follows from the expanded set of premises. As a solution, you introduce yet another premise: 'if the foregoing premises are true, the conclusion is true'. Regress. Hence: by this procedure you will never show that a conclusion follows from certain premises.
Source: What the Tortoise Said to Achilles

Russell's Relations

Suppose you want to reduce all relations. As a solution, you reduce them to properties of their relata. For example, you reduce the fact that a is earlier than b to the fact that a exists at t_1 and b exists at t_2. Yet, properties themselves stand in relations. As a solution, you reduce the latter in turn to properties of their relata. Regress. Hence: you will never reduce all relations.
Source: *Principles of Mathematics*, Sect. 214

McTaggart's Time

Suppose you want to resolve the contradictions involved in the A-series, i.e. series of items ordered temporally by the properties past, present and future (rather than temporal relations). As a solution, you appeal to a higher-order A-series. Yet, this involves contradictions as well. To resolve them, you appeal to a further A-series. Regress. Hence: you will never get rid of all contradictions involved in the A-series.
Source: The Unreality of Time

Wittgenstein's Rules

Suppose the correct use of '+2' is fixed, that the correct use of any linguistic or mental item is fixed by a rule, and that rules themselves are linguistic or mental items. This yields a regress that is absurd. Hence: it is not the case that the correct use of any linguistic or mental item is fixed by a rule.
Source: *Philosophical Investigations*, Sects. 185–224

Ryle's Knowledge-How

Suppose there are intelligent actions, that any action is intelligent only if the agent applies knowledge-that, and that applying knowledge-that is itself an intelligent action. This yields a regress that is absurd. Hence: it is not the case that any action is intelligent only if the agent applies knowledge-that.
Source: *The Concept of Mind*, Chap. 2, Knowing How and Knowing That

Tarski's Liar

Suppose you have to eliminate the liar paradox. As a solution, you hold that sentences (of the language L at issue) cannot speak of their own truth: this can only be stated in a metalanguage M. Surely you do not want M to be paradoxical. To ensure it is not, you appeal to yet another metalanguage. Regress. Hence: you will need an infinite hierarchy of languages.
Source: The Semantic Conception of Truth

1.4 Two Desiderata

A theory of IRAs should explain at least two things:

- For what purposes an IRA can be used;
- How an IRA should be evaluated.

For if we know these two things, we basically know what IRAs are and how they work. Comparable desiderata apply to theories about other common techniques of philosophical argumentation such as transcendental arguments and thought experiments.[3]

The first thing that should be explained by a theory of IRAs is the general goal of IRAs, i.e. the goal or goals shared by specific IRAs such as the classic ones listed in Sect. 1.3. What can one do with an IRA? It is often thought that an IRA can be employed as a negative, destructive weapon against one's opponent's views. For example, McTaggart employs an IRA in order to show that a certain theory of time (i.e. the so-called A-theory) is incorrect. But there are positive, constructive goals as well. For example, Aristotle uses an IRA in order to demonstrate the necessity of a highest good, i.e. something that is desired for the sake of itself.

Showing what the goals of IRAs are or can be, involves showing how such goals can be reached. That is, it involves showing what kind of reasoning steps are part of such arguments. Importantly, if infinite regresses are not formulated in the context of a broader argument, they are idle. For example, an infinite regress of

---

[3] For transcendental arguments, cf. Stern (2011). For thought experiments, cf. Sorensen (1992), Häggqvist (1996), Gendler (2000), Williamson (2007, Chap. 6). For introductions to philosophical methodology generally, cf. Rosenberg (1978), Baggini and Fosl (2003), Daly (2010).

reasons by itself does not show anything. It also needs to be shown how an infinite regress is generated in the first place (i.e. what assumptions are made), and what follows from such a regress (i.e. how a conclusion can be drawn from it).

A theory of IRAs should therefore provide a general template (or several such templates) that fixes the goal of an IRA and shows how that goal can be reached. For example, if the goal of an IRA is to refute a proposition because it entails an absurd regress, then a theory of IRAs should say what form such a proposition must take, how the regress is entailed, and how it finally follows that the proposition is refuted. As we will see, this basically means that the theory should provide *argument schemas* of which specific IRAs are an instance. Roughly, an argument schema is the general form of an argument where all variable parts are replaced with schematic letters. Consider the following example[4]:

(1) Socrates φ-s you.
(2) For all x, if Socrates φ-s x, then Socrates ψ-s x.
(3) If Socrates φ-s you, then Socrates ψ-s you. [from 2]
(4) Hence: Socrates ψ-s you. [from 1, 3]

Here (1) and (2) are the premises, and (4) is obtained from (1)–(3) on the basis of the rules Universal Instantiation and Modus Ponens. By systematically replacing the Greek letters 'φ' with 'teach' and 'ψ' with 'corrupt', we obtain the following instance of this schema:

(1) Socrates teaches you.
(2) For all x, if Socrates teaches x, then Socrates corrupts x.
(3) If Socrates teaches you, then Socrates corrupts you. [from 2]
(4) Hence: Socrates corrupts you. [from 1, 3]

Given that the schema was logically valid, and that this argument is an instance of it, this argument is logically valid too: the conclusion follows from the premises so that the former holds if the latter do. Now the first question is whether any such schemas are available for IRAs.

The second thing a theory of IRAs should explain is how a given IRA should be evaluated. Even if all IRAs can be logically valid arguments (i.e. if set out as instances of certain schemas), certainly not all IRAs are sound and have a true conclusion. For example, does it really follow from the regress of guardians that no-one is reliable? Or from the regress of criteria that no dispute can be settled? Hence the theory should indicate what premises have to be checked in order to determine whether an IRA is sound. Generally, all IRAs consist of the following components: a set of hypotheses and premises, which leads to an infinite regress, and a conclusion drawn from that regress. As a consequence, the evaluation of IRAs concerns two main issues:

---

[4] Throughout the book, the square brackets indicate how the line is obtained from previous lines.

- Is the suggested infinite regress indeed generated?
- If it does, then what is demonstrated by the regress?

Usually both steps involve controversy. In the case of the Problem of the Criterion it is controversial, first, whether the premises hold which together ensure that a whole series of disputes have to be settled, and second, what has been proven by this if it is agreed that they do.

Part of the second step, as we will see, consists in explaining when infinite regresses are *vicious*. It is generally accepted that certain infinite regresses are not vicious, i.e. entail no bad consequences. For example, many philosophers think that infinite regresses of causes are non-vicious (and hence that the Cosmological Argument for the existence of God is unsound; for discussion, cf. Martin 2007). Hence, the question arises as to how vicious regresses differ from non-vicious ones. Any adequate theory will enable a clear differentiation of these two.

These two steps are illustrated in some detail on the basis of two case studies in Chaps. 4 and 5. In Chap. 4 on Carroll's famous regress, I discuss when exactly regresses are generated and when not. In Chap. 5 on a fascinating case study from ethics, I discuss what a regress exactly proves if it is agreed upon that it is generated.

Two theories have been developed in the literature on IRAs. Both meet the foregoing two desiderata. Yet their explanations of how IRAs are to be used and evaluated are slightly different. These two theories will be called The *Paradox Theory* and The *Failure Theory*. These are my labels to structure the literature, and will be explained in due course.

Both theories are based on the idea that IRAs are pieces of *hypothetical reasoning*, i.e. arguments where some claims are considered, yet not taken to be true, for the sake of deriving absurd or other kinds of consequences from them.

According to the Paradox Theory, IRAs demonstrate that certain propositions are false because they have regressive consequences. Specifically, they are arguments where a certain claim X is considered, yet not taken to be true, for the sake of deriving a regressive consequence from it that conflicts with independent considerations, so that X has to be rejected by the hypothetical rule Reductio Ad Absurdum. This is briefly expressed by the following dictionary entry on IRAs:

> Since the existence of this regress is inconsistent with an obvious truth, we may conclude that the regress is vicious and consequently that the principle that generates it is false. (Tolhurst 1995, *Cambridge Dictionary of Philosophy*)

According to the Failure Theory, in contrast, IRAs demonstrate that certain solutions fail because they get stuck in a regress of problems that must be solved in order to solve the initial one. Specifically, they are arguments where a certain solution X to a given problem is considered, yet not taken to be true, for the sake of deriving a failure from it, so that 'if X, then failure' follows by the hypothetical rule Conditional Proof. This is briefly expressed by the following, different dictionary entry on IRAs:

A strategy gives rise to a vicious regress if whatever problem it was designed to solve remains as much in need of the same treatment after its use as before. (Blackburn 1994, *Oxford Dictionary of Philosophy*)

Thus, the Paradox and Failure Theories show that IRAs can be used for two main purposes: to demonstrate that certain propositions are false, or to demonstrate that certain solutions fail (and these negative outcomes can be associated, as we will see, with certain positive consequences). These theories are explained in Chaps. 2 and 3.

References

Amico, R.P. 1993. *The problem of the criterion*. Lanham: Rowman and Littlefield.

Aquinas, T. 1952. *Summa theologica*. Trans. Fathers of the English Dominican Province. Chicago: Encyclopaedia Britannica.

Aristotle. 1925. *Nicomachean ethics*. Trans. W.D. Ross. Oxford: Oxford University Press.

Baggini, J., and P.S. Fosl. 2003. *The philosopher's toolkit*. 2nd ed. 2010. Oxford: Blackwell.

Blackburn, S. 1994. Regress. *Oxford dictionary of philosophy*. 2nd ed. 2005. Oxford: Oxford University Press.

Bradley, F.H. 1893. *Appearance and reality*. 2nd ed. 1897. Oxford: Clarendon.

Carroll, L. 1895. What the tortoise said to Achilles. *Mind* 4: 278–280.

Chisholm, R.M. 1982. *The foundations of knowing*. Minneapolis: MUP.

Daly, C. 2010. *Introduction to philosophical methods*. Peterborough: Broadview.

Gendler, T.Z. 2000. *Thought experiment. On the powers and limits of imaginary cases*. New York: Garland.

Häggqvist, S. 1996. *Thought experiments in philosophy*. Stockholm: Almqvist and Wiksell.

Hume, D. 1748. *An enquiry concerning human understanding*, eds. Selby-Bigge L.A. and Nidditch P.H. 3rd ed. 1975. Oxford: Clarendon.

Hurwicz, L. 2008. But who will guard the guardians? *American Economic Review* 98: 577–585.

Juvenal. 1992. *The satires*. Trans. N. Rudd. Oxford: Oxford University Press.

Lammenranta, M. 2008. The Pyrrhonian problematic. In *The Oxford handbook of skepticism*, ed. J. Greco, 9–33. Oxford: Oxford University Press.

Martin, M. (ed.). 2007. *The Cambridge companion to atheism*. Cambridge: Cambridge University Press.

McTaggart, J.E. 1908. The unreality of time. *Mind* 17: 457–474.

Plato. 1996. *Parmenides*. Trans. M.L. Gill & P. Ryan. Indianapolis: Hackett.

Rosenberg, J.F. 1978. *The practice of philosophy*. Englewood Cliffs: Prentice-Hall.

Russell, B. 1903. *The principles of mathematics*. 2nd ed. 1937. London: Allen & Unwin.

Ryle, G. 1949. *The concept of mind*. Chicago: UCP.

Sextus Empiricus. 1996. *Outlines of Pyrrhonism*. Trans. B. Mates. *The skeptic way*. Oxford: Oxford University Press.

Sorensen, R.A. 1992. *Thought experiments*. Oxford: Oxford University Press.

Stern, R. 2011. Transcendental arguments. In *Stanford Encyclopedia of Philosophy*, ed. E.N. Zalta.

Tarski, A. 1944. The semantic conception of truth, and the foundations of semantics. *Philosophy and Phenomenological Research* 4: 341–376.

Tolhurst, W. 1995. Vicious regress. In *Cambridge dictionary of philosophy*, ed. R. Audi, 2nd ed. 1999. Cambridge: Cambridge University Press.

Wieland, J.W. 2013a. Infinite regress arguments. *Acta Analytica* 28: 95–109.

Wieland, J.W. 2013b. Strong and weak regress arguments. *Logique and Analyse* 224: 439–461.

Wieland, J.W. 2013c. What Carroll's tortoise actually proves. *Ethical Theory and Moral Practice* 16: 983–997.
Wieland, J.W. 2014. Access and the shirker problem. *American Philosophical Quarterly.*
Williamson, T. 2007. *The philosophy of philosophy.* Malden: Blackwell.
Wittgenstein, L. 1953. *Philosophical investigations.* Trans. G.E.M. Anscombe et al. 4th ed. 2009. Oxford: Blackwell.

Chapter 2
The Paradox Theory

In this chapter, it is explained for what purposes an IRA can be used, and how an IRA should be evaluated, according to the Paradox Theory of IRAs (one of the two theories discussed in this book).

2.1 Use

According to this theory, the goal of an IRA is to refute a proposition. An IRA can refute two kinds of propositions: namely, a universally quantified proposition (such as 'for all propositions x, x is justified to an agent S only if S has a reason for x') or an existentially quantified proposition (such as 'for at least proposition x, x is justified to an agent S'). These options will be explained in turn in the following. The main supporters of this theory are Black (1996) and Gratton (1997, 2009).[1]

The argument schema to refute universally quantified propositions may be represented as follows:

Paradox Schema A

(1) For all x in domain K, x is F only if there is a new item y in K and x and y stand in R.
(2) For all x and y in K, x and y stand in R only if y is F.
(3) At least one item in K is F.
(4) An infinity of items in K are F. [from 1–3]
(5) (4) is false: No infinity of items in K are F.
(C) (1) is false: It is not the case that for all x in K, x is F only if x stands in R to a new item y in K. [from 1–5]

---

[1] Versions of this theory have also been discussed or suggested, if only briefly, by Russell (1903, Sect. 329), Beth (1952), Yalden-Thomson (1964), Gettier (1965), Schlesinger (1983, Chap. 8), Sanford (1975, 1984), Day (1986, 1987), Clark (1988), Post (1993), Jacquette (1996), Nolan (2001), Klein (2003), Orilia (2006), Oppy (2006, Chap. 9), Maurin (2007, 2013), Cling (2008, 2009), Rescher (2010), Wieland (2012, 2013).

J. W. Wieland, *Infinite Regress Arguments*, SpringerBriefs in Philosophy,
DOI: 10.1007/978-3-319-06206-8_2, © The Author(s) 2014

This schema has one hypothesis for Reductio Ad Absurdum (RAA), i.e. line (1); three premises, i.e. lines (2), (3) and (5); and two main inferences, i.e. lines (4) and (C). For details about the inferences, see Sect. 2.4. To obtain instances of this schema, 'K' has to be replaced with a specific domain, and 'F' and 'R' with a predicate which expresses a property of or relation between the items in that domain. For example, this schema can be read on the basis of the following key[2]:

- domain K: persons;
- x is F: x is reliable;
- x and y stand in R: x is guarded by y.

These instructions yield the following instance of Paradox Schema A:

Guardians (Paradox A instance)

(1) For all persons x, x is reliable only if x is guarded by a guardian.
(2) For all persons x and y, x is guarded by a guardian y only if y is reliable.
(3) At least one person is reliable.
(4) An infinity of persons is reliable. [from 1–3]
(5) (4) is false: No infinity of persons is reliable.
(C) (1) is false: It is not the case that for all persons x, x is reliable only if x is guarded by a guardian. [from 1–5]

As a second example, here is the Problem of the Criterion (introduced in Sect. 1.2) constructed as such an instance:

Disputes (Paradox A instance)

(1) For all propositions x, the dispute about x is settled only if it is settled by a criterion.
(2) For all propositions x and y, the dispute about x is settled by y only if the dispute about y is settled.
(3) The dispute about at least one proposition is settled.
(4) The dispute about an infinity of propositions is settled. [from 1–3]
(5) (4) is false: It is not the case that the dispute about an infinity of propositions is settled.
(C) (1) is false: It is not the case that for all propositions x, the dispute about x is settled only if it is settled by a criterion. [from 1–5]

Such negative conclusions can be associated with certain positive outcomes, in these cases that there is at least one reliable person who is not guarded by a guardian, and that there is at least one dispute which is settled yet not by a criterion. Examples can easily be multiplied. For example, Aristotle's case reconstructed in terms of this schema would conclude that there is at least one thing that is good and not desired for the sake of something else.

---

[2] For an overview of further instances of the letters, see Sect. 2.3 below.

Next, the argument schema to refute existentially quantified propositions:

Paradox Schema B

(1) For all x in domain K, x is F only if there is a new item y in K and x and y stand in R.
(2) For all x and y in K, x and y stand in R only if y is F.
(3) At least one item in K is F.
(4) An infinity of items in K are F. [from 1–3]
(5) (4) is false: No infinity of items in K are F.
(C) (3) is false: No item in K is F. [from 1–5]

The only difference with the previous schema is that this time (3), rather than (1), is the hypothesis for RAA. In this case, the conclusions of the guardians and disputes IRAs would be respectively: no person is reliable; the dispute about no proposition is settled.

These two argument schemas are labelled 'Paradox Schemas' because their instances closely resemble paradoxes. Paradoxes (or at any rate many of them) are such that a set of propositions (which are to some extent independently intuitive) together entail a contradiction so that, by RAA, at least one of them must be false (cf. Sainsbury 1987, p. 1; Clark 2002, pp. 151–154). The same applies to instances of the Paradox Schemas: the propositions (1), (2) and (3) jointly entail (4) which forms a contradiction with (5) so that, by RAA, at least one of (1), (2), (3) or (5) must be false.

2.2 Evaluation

When are IRAs that take the form of the Paradox Schemas sound? Or put differently: how can one resist such arguments? To explain this, we need to consider the broader dialectical context of instances of the Paradox Schemas, i.e. a context between opponents who attack one other's position and defend their own. If we consider two persons 'S1' and 'S2', then the dialectical situation is as follows:

Line	What	Dialectical context
(1)/(3)	Hypothesis for RAA	S1's position
(2)	Premise	S2 shows that S1 has to concede this
(1)/(3)	Premise	S2 shows that S1 has to concede this
(4)	Infinite regress	S2 infers this from (1) to (3)
(5)	Premise	S2 shows that S1 has to concede this
(C)	Rejection	S2 infers this from (1) to (5) by RAA

The slash '/' indicates the difference between the two Paradox Schemas.

Consider for example the following IRA[3]:

Reasons (Paradox B instance)

(1) For all propositions x, x is justified to S only if S has a reason y for x.
(2) For all propositions x and y, S has a reason y for x only if y is justified to S.
(3) At least one proposition is justified to S.
(4) S has an infinity of reasons. [from 1–3]
(5) (4) is false: S does not have an infinity of reasons.
(C) (3) is false: No proposition is justified to S. [from 1–5]

In this case, the dialectic is between epistemological views that hold that justification in fact obtains (S1) and the sceptical position that no proposition is justified to anyone (S2). First, S1's position is constructed as the view that at least one proposition is justified to someone. Second, S2 defends that a proposition is justified to someone only if she has a reason for it. Third, S2 defends that someone has a reason only if that reason itself is justified to her. Fourth, S2 infers an infinite regress from the foregoing. Fifth, S2 defends that the regress is absurd or otherwise vicious (so that (5) is true: it is true that one does not have an infinity of reasons). Last, S2 rejects S1's position and concludes that no proposition is justified to anyone.

From this, it can easily be seen what can be done to resist an instance of a Paradox Schema. There are five main options, corresponding to each of the lines. Person S1 could deny that:

- the hypothesis was in fact her position;
- the first premise that helps generating the regress holds;
- the second premise that helps generating the regress holds;
- an infinite regress is entailed even if the foregoing does hold;
- the regress does not exist (or is unacceptable in another way).

In the regress of reasons case, for example, a popular option is the second. Namely (a version of) foundationalism denies that reasons are always required for justification. The last option goes under the name 'infinitism', and denies that the regress of reasons is unacceptable in the first place.

Indeed, not all infinite regresses are thought to be vicious or unacceptable. There are 'good' and 'bad' cases. What is the difference? To explain this, we need the concept of an infinite regress in general. According to the Paradox Theory, all infinite regresses are entailed by the schematic lines (1)–(3) of the Paradox Schemas and consist of steps each of which is a necessary condition for the previous one. Schematically:

---

[3] As the focus here is on IRAs generally, I will ignore some details regarding the content of this instance, such as the difference between propositional and doxastic justification (cf. Klein 2007), or the difference between justification as a state and justification as an activity (cf. Rescorla 2014).

(a) a is F;
(b) a and b stand in R;
(c) b is F;
(d) b and c stand in R;
etc.

Here, for example, are the regresses of reasons and guardians in Paradox format (where p_{1-n} are names for propositions and persons respectively):

(a) p_1 is justified to S;
(b) S has a reason p_2 for p_1;
(c) p_2 is justified to S;
(d) S has a reason p_3 for p_2;
etc.

(a) p_1 is reliable;
(b) p_1 is guarded by a guardian p_2;
(c) p_2 is reliable;
(d) p_2 is guarded by a guardian p_3;
etc.

In series like these, (b) is a necessary condition for (a), (c) is a necessary condition for (b), and so on. Importantly, not just any series of the form (a), (b), (c), etc. is a regress. For example, a mere series of guardians that guard one another is not a regress. Likewise, a mere series of numbers (1, 2, 3, etc.) is not a regress. Only those series entailed by the schematic lines (1)–(3) of the Paradox Schemas are considered as regresses.

Furthermore, an infinite regress is vicious iff the entailed infinite regress does not exist (or is shown to be unacceptable, as I will explain below). For in that case is one committed to a contradiction (i.e. between lines (4) and (5) of the Paradox Schemas: an infinity of Ks are F and no infinity of Ks are F), and must one reject one of the propositions that generates the infinite regress.

For example, infinite regresses of reasons are vicious only if it has been shown that there are no such regresses, i.e. that the infinity of necessary conditions are not, or cannot be, in place (e.g. that one does not or cannot have an infinity of reasons). For example, one worry would be that it is mentally impossible for human beings to possess so many of reasons. Furthermore, such regresses are non-vicious if it has been shown that they can and do exist. As noted, the view that they are not vicious is called 'infinitism'.[4] In Sect. 5.3, I will provide one extended example of a regress (generated in a Paradox way) that is arguably harmless.

One general reason why regresses are vicious concerns paradoxes of infinity (for an overview, cf. Oppy 2006, Chap. 3). For example, in the case of Hilbert's Hotel with an infinite number of rooms, all of which are occupied, the question is

---

[4] For defences, cf. Klein (1999, 2007), Peijnenburg (2010), Aikin (2011).

how it can be that there is always room for a new guest (namely by moving the guest from room 1 to room 2, the guest from 2 to 3, and so on). Moreover, if infinities are regarded as paradoxical and absurd from the very start, then all regresses (generated in a Paradox way) are vicious. For surely if they are impossible, they do not exist.

In many cases, however, it is generally accepted that the regress is possible, i.e. that it could exist, yet debated whether it is *acceptable*, i.e. whether the benefits of rejecting one of the propositions that generate the regress outweigh the costs of the regress (cf. Lewis 1983, pp. 353–354; Nolan 2001; Cameron 2008). Consider for example the regressive claim that for every x, there is a singleton {x} (i.e. the set of itself). This generates regresses such as: Socrates, {Socrates}, {{Socrates}}, and so on. The question is whether the benefits of rejecting the claim that anything has a singleton outweigh the costs of such regresses (in this case infinitely many sets). The regress of singletons is vicious, then, only if the costs are too high.

2.3 Classic Instances

Here is a list of further filling instructions for the Paradox Schemas (based on the classic cases from Sect. 1.3). In each case, to be sure, further details of the schematic letters could be spelled out. Yet this would go beyond what theories of IRAs have to provide: namely, the form that IRAs (or a group among them) have in common.

Items in K	x is F	x and y stand in R
Sets of large things	the members of x are large	y contains the form Largeness in which the members of x participate
Actions	x is good	y is an end for the sake of which x is performed
Propositions	x is justified to S	y is a reason for x that is available to S
Contingent beings	x exists	x is caused by y
Inductive inferences	x is justified	x is derived from past facts and the assumption y that the future resembles the past
Relations	x is unified with its relata	y unifies x with its relata
Sets of premises	a conclusion follows from x	y contains an additional premise 'if the members of x are true, then the conclusion must be true'
Sets of relata	the members of x stand in an asymmetric relation	y is a set of properties such that the members of x have these properties
A-series	x is contradiction-free	y is an A-series such that the members of x are past, present and future at different times of y
Rules	x has a fixed use	y fixes the use of x
Actions	x is performed intelligently	y is an action of applying knowledge that x must be performed in such and such way
Languages	x is liar paradox-free	that sentences of x are true or not is only stated in a meta-language y

2.4 Logical Analysis

In this section, I provide the logical details of the Paradox Schemas presented in Sect. 2.1. It shows that they are valid according to classical first-order logic. A few comments are in order. I use the propositional calculus by Nolt et al. (1988, Chap. 4), and the first-order extension by Gamut (1982, pp. 142–147). This means that I employ standard natural deduction abbreviations of the inference rules, a strict distinction between premises (PREM) and hypotheses (HYP), and the hypothetical rules Reductio Ad Absurdum ($\neg$I) and Conditional Proof ($\rightarrow$I). All portions of hypothetical reasoning are clearly marked by vertical lines. Some of the predicates and premises need some explanation. These explanations are provided right after the formalisation.

Key:

Kx: x is in domain K
Fx: x has property F
Rxy: x stands in relation R to y

Example:

Kx: x is a proposition
Fx: the dispute about x is settled
Rxy: the dispute about x is settled by y

Paradox Schema A

(1)	$\forall x\forall y((Ky\wedge Rxy)\rightarrow Fy)$	PREM
(2)	$\exists x(Kx\wedge Fx)$	PREM
(3)	$\neg(\exists x(Kx\wedge Fx)\wedge(\forall x((Kx\wedge Fx)\rightarrow\exists y(Ky\wedge Fy\wedge Rxy))))$	PREM
(4)	$\forall x((Kx\wedge Fx)\rightarrow\exists y(Ky\wedge Rxy))$	HYP $\neg$I
(5)	$Ka\wedge Fa$	HYP $\rightarrow$I
(6)	$(Ka\wedge Fa)\rightarrow\exists y(Ky\wedge Ray)$	4; $\forall$E
(7)	$\exists y(Ky\wedge Ray)$	5, 6; $\rightarrow$E
(8)	$Kb\wedge Rab$	HYP $\rightarrow$I
(9)	$\forall x((Kb\wedge Rxb)\rightarrow Fb)$	1; $\forall$E
(10)	$(Kb\wedge Rab)\rightarrow Fb$	9; $\forall$E
(11)	Fb	8, 10; $\rightarrow$E
(12)	$Kb\wedge Fb\wedge Rab$	8, 11; $\wedge$I
(13)	$\exists y(Ky\wedge Fy\wedge Ray)$	12; $\exists$I
(14)	$Kb\wedge Rab\rightarrow\exists y(Ky\wedge Fy\wedge Ray)$	8-13; $\rightarrow$I
(15)	$\exists y(Ky\wedge Fy\wedge Ray)$	7, 14; $\exists$E
(16)	$(Ka\wedge Fa)\rightarrow\exists y(Ky\wedge Fy\wedge Ray)$	5-15; $\rightarrow$I
(17)	$\forall x((Kx\wedge Fx)\rightarrow\exists y(Ky\wedge Fy\wedge Rxy))$	16; $\forall$I
(18)	$\exists x(Kx\wedge Fx)\wedge(\forall x((Kx\wedge Fx)\rightarrow\exists y(Ky\wedge Fy\wedge Rxy)))$	2, 17; $\wedge$I
(19)	$(18)\wedge\neg(18)$	18, 3; $\wedge$I
(20)	$\neg(\forall x((Kx\wedge Fx)\rightarrow\exists y(Ky\wedge Rxy)))$	4-19; $\neg$I

Lines (1)–(4) may have variants in terms of one- or many-place predicates and their number (this does not hold for the Failure Schemas). Also, it is easy to see how variant B of the Paradox Schema can be constructed where line (2), rather than (4), is the hypothesis for Reductio Ad Absurdum ($\neg$I).

Line (18) requires some explanation. Literally, it does not yet express that there is an infinity of Ks that are F. The reason is that the existential quantifier does not yet say what it should say, namely that 'y' has to be a new item in the domain. The phrase 'there is a new item y' cannot be expressed by a familiar logical constant, for it does not mean merely 'there is an item y that is distinct from x', but rather 'there is an item y that is distinct from *all other items mentioned earlier in the regress*'. To express this, we could introduce an additional relation '<', distinct from R, whose only job is to order the Ks, and make sure that all items introduced in the regress are new items (so that they form an infinite, non-circular series). To do this, 'x < y' can be read as 'x occurs earlier in the regress than y' and has to satisfy the following conditions[5]:

- $\forall x \neg x{<}x$
- $\forall x \forall y \forall z ((x{<}y \wedge y{<}z) \rightarrow x{<}z)$
- $\forall x \forall y ((x{\neq}y \wedge Kx \wedge Ky) \rightarrow (x{<}y \vee y{<}x))$
- $\forall x \forall y (x{<}y \rightarrow (Kx \wedge Ky))$

Moreover, this allows us to formulate the contradiction in (19) between 'an infinity of Ks are F' and 'no infinity of Ks are F' in first-order terms:

- $\exists x (Kx \wedge Fx) \wedge \forall x ((Kx \wedge Fx) \rightarrow \exists y (x{<}y \wedge Ky \wedge Fy))$
- $\exists x (Kx \wedge Fx \wedge \forall y ((Ky \wedge x{<}y) \rightarrow \neg Fy))$

For example: The dispute about at least one proposition is settled and the dispute about any proposition is settled only if there is a new proposition about which the dispute is settled; For at least one proposition x, the dispute about x is settled, and for all new propositions y, the dispute about y is not settled.

Another option, suggested by Cling (2009, p. 343), would be to drop the idea of 'infinity', and replace 'there is an infinity of Ks that are F' with 'there is an endless regress of Ks that are F' (where the latter, but not the former, includes finite, circular regresses). If we change this throughout the argument we would not need to block loops, and yet we would still obtain a contradiction in (19) so that we can apply $\neg$I. This solution will work in all cases where infinity is not really an issue (i.e. where the unacceptability of a regress does not derive from its infinity).

Finally: a very similar logical analysis can be provided for Paradox Schema B (i.e. which differs mainly regarding HYP $\neg$I).

---

[5] These ensure that '<' is irreflexive and transitive, and that all and only Ks stand in '<'. Thanks to Christian Straßer for suggesting this solution.

References

Aikin, S.F. 2011. *Epistemology and the regress problem*. New York: Routledge.

Beth, E.W. 1952. The prehistory of research into foundations. *British Journal for the Philosophy of Science* 3: 58–81.

Black, O. 1996. Infinite regress arguments and infinite regresses. *Acta Analytica* 16: 95–124.

Cameron, R. 2008. Turtles all the way down: regress, priority and fundamentality. *Philosophical Quarterly* 58: 1–14.

Clark, M. 2002. *Paradoxes from A to Z*. 2nd ed. 2007. London: Routledge.

Clark, R. 1988. Vicious infinite regress arguments. *Philosophical Perspectives* 2: 369–380.

Cling, A.D. 2008. The epistemic regress problem. *Philosophical Studies* 140: 401–421.

Cling, A.D. 2009. Reasons, regresses and tragedy: the epistemic regress problem and the problem of the criterion. *American Philosophical Quarterly* 46: 333–346.

Day, T.J. 1986. *Infinite regress arguments. Some metaphysical and epistemological problems*. Ph.D. dissertation, Indiana University.

Day, T.J. 1987. Infinite regress arguments. *Philosophical Papers* 16: 155–164.

Gamut, L.T.F. 1982. Introduction to logic. In *Logic, language and meaning*. Vol. 1, Trans. 1991. Chicago: UCP.

Gettier, E.L. 1965. Review of Passmore's philosophical reasoning. *Philosophical Review* 2: 266–269.

Gratton, C. 1997. What is an infinite regress argument? *Informal Logic* 18: 203–224.

Gratton, C. 2009. *Infinite regress arguments*. Dordrecht: Springer.

Jacquette, D. 1996. Adversus adversus regressum (against infinite regress objections). *Journal of Speculative Philosophy* 10: 92–104.

Klein, P.D. 1999. Human knowledge and the infinite regress of reasons. *Philosophical Perspectives* 13: 297–325.

Klein, P.D. 2003. When infinite regresses are not vicious. *Philosophy and Phenomenological Research* 66: 718–729.

Klein, P.D. 2007. Human knowledge and the infinite progress of reasoning. *Philosophical Studies* 134: 1–17.

Lewis, D.K. 1983. New work for a theory of universals. *Australasian Journal of Philosophy* 61: 343–377.

Maurin, A.-S. 2007. Infinite regress: virtue or vice? In *Hommage à Wlodek*, eds. T. Rønnow-Rasmussen et al., 1–26. Lund University.

Maurin, A.-S. 2013. Infinite regress arguments. In *Johanssonian investigations*, eds. C. Svennerlind et al., 421–438. Heusenstamm: Ontos.

Nolan, D. 2001. What's wrong with infinite regresses? *Metaphilosophy* 32: 523–538.

Nolt, J., D. Rohatyn and A. Varzi 1988. *Theory and problems of logic*. 2nd ed. 1998. New York: McGraw-Hill.

Oppy, G. 2006. *Philosophical perspectives on infinity*. Cambridge: CUP.

Orilia, F. 2006. States of affairs: Bradley vs. Meinong. *Meinong Studies* 2: 213–238.

Peijnenburg, J. 2010. Ineffectual foundations. *Mind* 119: 1125–1133.

Post, J.F. 1993. Infinite regress argument. In *A companion to epistemology*, eds. J. Dancy et al., 2nd ed. 2010, 447–450. Oxford: Blackwell.

Rescher, N. 2010. *Infinite regress. The theory and history of a prominent mode of philosophical argumentation*. New Brunswick: Transaction.

Rescorla, M. 2014. Can perception halt the regress of justifications? In *Ad infinitum. New essays on epistemological infinitism*, eds. J. Turri and P. D. Klein, chap. 10. Oxford: OUP.

Russell, B. 1903. *The principles of mathematics*. 2nd ed. 1937. London: Allen & Unwin.

Sainsbury, R.M. 1987. *Paradoxes*. 3rd ed. 2009. Cambridge: CUP.

Sanford, D.H. 1975. Infinity and vagueness. *Philosophical Review* 84: 520–535.

Sanford, D.H. 1984. Infinite regress arguments. In *Principles of philosophical reasoning*, ed. J.H. Fetzer, 93–117. Totowa: Rowman & Allanheld.

Schlesinger, G.N. 1983. *Metaphysics. Methods and problems*. Oxford: Blackwell.
Wieland, J.W. 2012. *And so on. Two theories of regress arguments in philosophy*. Ph.D. dissertation, Ghent University.
Wieland, J.W. 2013. Infinite regress arguments. *Acta Analytica* 28: 95–109.
Yalden-Thomson, D.C. 1964. Remarks about philosophical refutations. *Monist* 48: 501–512.

Chapter 3
The Failure Theory

In this chapter, it is explained for what purposes an IRA can be used, and how an IRA should be evaluated, according to the Failure Theory of IRAs (one of the two theories discussed in this book).

3.1 Use

According to this theory, the goal of an IRA is to show that a solution fails to solve a given problem. For example, IRAs can demonstrate that a solution to define, analyse, explain, resolve, perform, etc. something fail (hence the name of the theory). These problem solving failures come in two kinds: either the solution fails to solve a universally quantified problem (such as the problem to justify all propositions) or it fails to solve an existentially quantified problem (such as the problem to justify at least one proposition). In the following these options will be explained in turn. The main supporters of this theory are Passmore (1961, Chap. 2) and Wieland (2012, 2013a, b).[1]

The argument schema to show that a solution fails to solve a universally quantified problem can be represented as follows:

Failure Schema A

(1) For all x in domain K, if S has to φ x, then S ψ-s x.
(2) For all x in K, if S ψ-s x, then there is a new item y in K.
(3) For all x in K, S has to φ x.
(4) For all x in K, if S has to φ x, then S has to φ a new item y in K. [from 1–3]
(5) S will never φ all items in K. [from 4]

---

[1] Versions of this theory have also been discussed or suggested, if only briefly, by Russell (1903, Sect. 329), Rankin (1969), Armstrong (1974, 1978), Johnson (1978, pp. 80–81), Rosenberg (1978, pp. 69–73), Schlesinger (1983, Chap. 8), Sanford (1984), Day (1986, 1987), Ruben (1990, pp. 125–129), Johnstone (1996), Rodriguez-Pereyra (2002, pp. 107–108), Gillett (2003), Maurin (2007, 2013), Rescher (2010), Bliss (2013).

J. W. Wieland, *Infinite Regress Arguments*, SpringerBriefs in Philosophy, DOI: 10.1007/978-3-319-06206-8_3, © The Author(s) 2014

(C) If S ψ-s any item in K that S has to φ, then S will never φ all items in K. [from 1–5]

This schema has one hypothesis for Conditional Proof (CP), i.e. line (1) (this is the considered solution for the problem at hand); two premises, i.e. lines (2) and (3); and three main inferences, i.e. lines (4) (5) and (C). For details about the inferences, see Sect. 3.5. To obtain instances of the schemas, 'K' has to be replaced with a specific domain, 'S' with a person (or agent that can solve problems), and the Greek letters 'φ' and 'ψ' with predicates which express actions involving the items in that domain. For example, this schema can be read on the basis of the following key[2]:

- S: you
- domain K: disputes;
- φ x: settle x;
- ψ x: invoke a criterion to settle x.

These instructions yield the following instance of Failure Schema A:

Disputes (Failure A instance)

(1) For all disputes x, if you have to settle x, then you invoke a criterion to settle x.
(2) For all disputes x, if you invoke a criterion to settle x, then there is a new dispute y about that criterion.
(3) For all disputes x, you have to settle x.
(4) For all disputes x, if you have to settle x, then you have to settle a new dispute y. [from 1–3]
(5) You will never settle all disputes. [from 4]
(C) If you invoke a criterion every time you have to settle a dispute, then you will never settle all disputes. [from 1–5]

Next, here is the argument schema to show that a solution fails to solve an existentially quantified problem:

Failure Schema B

(1) For all x in domain K, if S has to φ x, then S ψ-s x.
(2) For all x in K, if S ψ-s x, then there is a new item y in K and S first has to φ y in order to φ x.
(3) For all x in K, if S has to φ x, then there is a new item y in K and S first has to φ y in order to φ x. [from 1–2]
(4) S will never φ any item in K. [from 3]
(C) If S ψ-s any item in K that S has to φ, then S will never φ any item in K. [from 1–4]

---

[2] For an overview of further instances of the letters, see Sect. 3.3.

Whereas the rationale of Failure Schema A is that a universally quantified problem (i.e. of φ-ing all items in K) will never be solved because the solution under consideration generates a regress of further instances of this problem, the rationale of Failure Schema B is that no existentially quantified problem (i.e. of φ-ing at least one item x in K) will ever be solved, as the solution under consideration generates a regress of more and more problems that must be solved in order for any initial one to be solved.

What accounts for this difference? The difference is that line (2) of Failure B is more substantive: it not only states that new problems are generated, but also that the newly generated problems must be solved in order to solve any initial one in the first place. As a consequence, this premise suffices to generate a regress, i.e. without the further premise (3) of Failure Schema A.

Schema B can for example be read on the basis of the following key:

- S: you
- domain K: persons;
- φ x: have x guarded;
- ψ x: hire a guardian for x.

These instructions yield the following instance:

Guardians (Failure B instance)

(1) For all persons x, if you should have x guarded, then you hire a guardian for x.
(2) For all persons x, if you hire a guardian y for x, then you should have y guarded first in order to have x guarded.
(3) For all persons x, if you should have x guarded, then you should have another person guarded first in order to have x guarded. [from 1–2]
(4) You will never have anyone guarded. [from 3]
(5) If you hire a guardian every time you should have someone guarded, then you will never have anyone guarded. [from 1–4]

It is important to see that the Problem of the Criterion, set out earlier as an instance of Failure Schema A, can take the form of Schema B as well:

Disputes (Failure B instance)

(1) For all disputes x, if you have to settle x, then you invoke a criterion to settle x.
(2) For all disputes x, if you invoke a criterion to settle x, then you first have to settle a new dispute y about that criterion in order to settle x.
(3) For all disputes x, if you have to settle x, then you first have to settle a new dispute y about that criterion in order to settle x. [from 1–2]
(4) You will never settle any dispute. [from 3]
(C) If you invoke a criterion every time you have to settle a dispute, then you will never settle any dispute. [from 1–4]

This time the conclusion is that you fail to settle *any* dispute, rather than merely fail to settle *all* of them.

Conclusions which take the form of the Failure Schema (A or B) are interesting in at least the following two contexts. First, suppose that there is an alternative regress-free solution to the given problem (perhaps there are other solutions to have your partner guarded, to settle a dispute, to justify a proposition, to demonstrate that a conclusion follows, to reduce relations, to resolve paradoxes, etc.). In such cases, an IRA can be used, positively, to favour that alternative solution. Second, suppose that there is no such alternative solution to a given problem. In that case, the IRA can be used, negatively, to demonstrate that the problem cannot be solved (that partners cannot be guarded, disputes not settled, propositions not justified, etc.).

3.2 Evaluation

When are IRAs that take the form of the Failure Schemas sound? We will again explain this in terms of a broader dialectical context with two persons S1 and S2:

Line	What	Dialectical context
(1)	Solution for CP	S1's proposal
(2)/(3)	Premise(s)	S2 shows that S1 has to concede these
(3)/(4)	Infinite regress	S2 infers this from (1) to (2)/(3)
(4)/(5)	Failure	S2 infers this from (3)/(4)
(C)	If (2), then (4)/(5)	S2 infers this from (1) to (4)/(5) by CP

The slash '/' indicates the difference between the two Failure Schemas

Consider for example the following IRA[3]:

Reasons (Failure B instance)

(1) For all propositions x, if you have to justify x, then you provide a reason for x.
(2) For all propositions x, if you provide a reason y for x, then you first have to justify y in order to justify x.
(3) For all propositions x, if you have to justify x, then you first have to justify y in order to justify x. [from 1–2]
(4) You will never justify any proposition. [from 3]
(C) If you provide a reason for any proposition that you have to justify, then you will never justify any proposition. [from 1–4]

In this case, the dialectic is between non-sceptical views which hold that in order to justify a proposition one must appeal to reasons for them (S1) and the sceptical position that one will never justify anything via this strategy (S2). First,

---

[3] For reasons explained in Chap. 2, I will ignore some details regarding this instance.

S1's solution to the problem of justifying a proposition is to invoke a reason for it. Second, S2 shows that S1 does not solve any such problem in this way unless she justifies that reason as well. Third, S2 infers an infinite regress from the foregoing. Finally, S2 concludes that S1's attempt to justify at least one proposition fails.

As noted above, S2 has two options at this point: either she may use the argument in favour of an alternative regress-free solution, or she may conclude that the problem remains unsolved (the sceptical position).

From this it can easily be seen how to resist an instance of a Failure Schema. There are four main options, corresponding to each of the lines. Person S1 could deny that:

- her solution was in fact the one she proposed;
- the extra premise(s) that entail the regress hold(s);
- an infinite regress is entailed even if the foregoing is in place;
- the regress entails a failure.

In the regress of reasons case, for example, a popular option is the first. That is (a version of) foundationalism suggests that (1) does not hold unrestrictedly: for some propositions (namely the basic ones), one does not appeal to a reason in order to justify them.

Again, not all infinite regresses are thought to be vicious or unacceptable. To see this, consider the concept of an infinite regress generally. According to this theory, all infinite regresses are entailed by the schematic lines (1)–(2)/(3) of the Failure Schemas and consist of steps where each is either a problem, or a solution to the previous step. Schematically:

(a) S has to φ a;
(b) S ψ-s a;
(c) S has to φ b;
(d) S ψ-s b;
etc.

Here, for example, are the regresses of reasons and guardians in Failure format (where p_{1-n} are names for propositions and persons respectively):

(a) S has to justify p_1;
(b) S provides a reason p_2 for p_1;
(c) S has to justify p_2;
(d) S provides a reason p_3 for p_2;
etc.

(a) S should have p_1 guarded;
(b) S hires a guardian p_2 for p_1;
(c) S should have p_2 guarded;
(d) S hires a guardian p_3 for p_2;
etc.

In series like these, (a) is a problem, (b) is a solution to (a), (c) is a problem (entailed by the solution (b)), (d) is solution for (c), and so on. As in the previous theory, not just any series of the form (a), (b), (c), etc. is a regress. Only those series entailed by the schematic lines (1)–(2)/(3) of the Failure Schemas are considered as regresses.

Furthermore, an infinite regress is vicious iff it entails a failure, i.e. demonstrates that the solution that generates the regress will never solve the given problem.

Now, infinite regresses generated within the context of the Failure Schemas are almost always vicious. The reason is that there is no extra, substantial premise after the regress that can fail to be true. Recall: in the case of Paradox arguments, one has to show that the regress is absurd, or unacceptable in another way, in order to obtain a contradiction (as we saw in Chap. 2). No such step is needed in the case of Failure arguments. For example, in the IRA concerning reasons just given there is no substantial premise needed to obtain (4) from (3), i.e. to infer that S will never justify any proposition given that each time she has to justify another proposition first in order to justify any proposition.

In the Failure B case generally, from the fact that it is always the case that S has to solve another problem of a certain kind first in order to solve any such problem, we may immediately conclude that she never solves *any* problem of that kind. Similarly, in the Failure A case generally, from the fact that S always has to solve yet another problem of a certain kind, we may immediately conclude that she never solves *all* problems of that kind.

Nevertheless, in some selected cases regresses generated in the Failure way may still be non-vicious. Consider for example the following Zeno paradox[4]:

Achilles and the Tortoise (Failure A instance)

(1) For all distances to the Tortoise x, if Achilles has to traverse x, then he runs x.
(2) For all distances to the Tortoise x, if Achilles runs x, then there is a new distance y to the Tortoise.
(3) For all distances to the Tortoise x, Achilles has to traverse x.
(4) For all distances to the Tortoise x, if Achilles has to traverse x, then he has to traverse a new distance to the Tortoise y. [from 1–3]
(5) Achilles will never traverse all distances to the Tortoise. [from 4]
(C) If Achilles runs all distances to the Tortoise that he has to traverse, then he will never traverse all distances to the Tortoise. [from 1–5]

Line (4) entails a *supertask*: from the fact that Achilles always has to traverse yet another distance to the Tortoise it follows that Achilles has to traverse an

----

[4] The Zeno paradoxes can be found in Aristotle's *Physics*, 239b. For an overview, cf. Huggett (2002). Interestingly, Zeno's Dichotomy Paradox involves the very same instances of the schematic letters, yet has the form of Failure Schema B, rather than A. Its conclusion is 'if Achilles runs all distances to the Tortoise that he has to traverse, then he will never traverse *any* distance the Tortoise.'

infinity of distances to the Tortoise. Still, to conclude from this that Achilles will never succeed, i.e. will fail to traverse all distances to the Tortoise, is no doubt counterintuitive. Why would Achilles fail to catch the Tortoise by running to the latter? Achilles is much faster and will certainly catch the Tortoise eventually. Yet, if (5) fails to follow from (4), then the suppressed premise that licences the step from (4) to (5) must be false, and hence the regress non-vicious.[5] In Sect. 5.4, I will provide one extended and contemporary example of a regress (generated in a Failure way) that is arguably harmless.

3.3 Classic Instances

Here is a list of further filling instructions for the Failure Schemas (based on the classic cases from Sect. 1.3). In each case further details of the schematic letters could be given, yet, as noted in Sect. 2.3, that is not my task here.

Items in K	$\varphi\,x$	$\psi\,x$
Things	explain why x is large	refer to the fact that x partakes of the form Largeness
Actions	explain why x is good	refer to the fact that x is done for the sake of something else
Propositions	justify x	provide a reason for x
Contingent beings	explain why x exists	refer to x's cause
Inductive inferences	justify x	assume that the future resembles the past such that x can be derived from past facts
Relations	explain how x is unified with its relata	invoke an additional relation that unifies x with its relata
Premises	demonstrate that x entails a conclusion	introduce an extra premise 'if x is true, the conclusion must be true'
Relations	reduce x	posit corresponding properties of x's relata
A-series	resolve the contradictions involved with x	introduce a higher-order A-series to resolve the contradictions involved with x
Rules	fix the correct use of x	provide a more elaborate rule to fix the correct use of x
Actions	perform x intelligently	apply knowledge that x must be performed in such and such way
Languages	resolve the Liar Paradox in x	hold that no sentence in x can speak of its own truth

---

[5] For this premise, see Sect. 3.5.

3.4 Concluding Remarks

The two theories of IRAs explained in the foregoing each provide two argument schemas. This makes for four different possible uses an IRA could have. They can be used:

- To refute a universally quantified proposition of the form 'any item in a certain domain is F only if it stands in R to another item in that domain';
- To refute an existentially quantified proposition of the form 'at least one item in the given domain is F';
- To show that a solution fails to solve a universally quantified problem of the form 'having to φ all items in the given domain';
- To show that a solution fails to solve an existentially quantified problem of the form 'having to φ at least one item in the given domain'.

The main difference between the two theories lies in the step from regress to conclusion. If you want to conclude that a certain proposition is false, then, as the Paradox Schemas make clear, you should come with independent considerations that the regress does not exist. In contrast, if you want to conclude that a certain solution fails, then, as the Failure Schemas make clear, no such considerations are needed.[6] The importance of this difference will be illustrated in some detail in Chap. 5.

At this point, two general questions arise:

- If you want to evaluate an existing IRA, then which argument schema should you invoke?
- If you would like to make an IRA yourself, then how should you proceed?

Let me take these in turn. As to the first question, at least two rules guide argument evaluation (cf. Feldman 1993, p. 115):

Interpretation. If one wants to evaluate an IRA, then one should capture the actual claims in a given debate.

Charity. If one wants to evaluate an IRA, then one should address the argument in its most interesting format (i.e. with the most plausible premises and strongest conclusion).

Consider for example the IRA of disputes. If we have to choose between evaluating it as an instance of Failure Schema A (with a conclusion of the form 'if S employs solution such and such, then S will never settle *all* disputes') or as an instance of Failure Schema B (with a conclusion of the form 'if S employs solution such and such, then S will never settle *at least one* dispute'), then we may choose the latter on the basis of both Interpretation and Charity. As to the Interpretation

---

[6] For further discussion of these differences, see in particular Black (1996) and Wieland (2012, 2013a).

rule, it is closer to what the ancient sceptics had in mind, at least according to Sextus Empiricus. Specifically, a global suspension of belief follows only if it is impossible to settle disputes (about beliefs) generally. As to Charity, the B format is the more interesting of the two formats as it has plausible premises and the strongest conclusion. Namely, the B-conclusion entails the A-conclusion: if S will never settle any dispute, then S will necessarily never settle all of them (but not vice versa).

Yet, certain other cases such as Plato's Third Man, McTaggart's attack on the A-theory, and Tarski's response to the Liar Paradox are more plausible in Failure A format. The same holds for the following everyday example (from Armstrong 1978, p. 21):

> Suppose you want to get rid of the bulge in the carpet. As a solution, you press it down. Yet, as happens with bulges in carpets, it reappears elsewhere in the carpet. So a similar problem occurs: there is a bulge in the carpet, and you want to get rid of it. As a solution, you press down more bulges. Regress.

Clearly, the conclusion of this regress is not that you will never press down *any* bulge, merely that you will never press down *all* bulges. Hence, this case should be reconstructed in Failure A format.

This leads me to the second issue: how to make an IRA yourself? Making an IRA is not difficult with the argument schemas at our disposal: you just have to select one of the four possible goals identified above, and fill out the schematic letters of the corresponding schema. Clearly, however, making an IRA that is sound and has true premises is slightly more difficult. Here is my own attempt at constructing a sound, everyday example (and I will leave it to the reader to spell out the exact argument):

> Suppose you want to get rid of your anger. As a solution, you kill one of your neighbour's cows. Yet, as happens with neighbours, she gets angry and kills one of your cows. So a similar problem occurs: your neighbour killed one of your cows and you want to get rid of your anger. As a solution, you kill another of her cows. Regress. Hence, taking revenge is a bad way to get rid of your anger.

Surely in each case argument evaluation and construction remains a delicate enterprise, but the meta-debate on IRAs has at least delivered four specific options as to what forms an IRA can possibly take.

3.5 Logical Analysis

In this section, I provide the logical details of the Failure Schemas presented in Sect. 3.1. It shows that they are valid according to classical first-order logic. A few comments are in order. I use the propositional calculus by Nolt et al. (1988, Chap. 4), and the first-order extension by Gamut (1982, pp. 142–147). This means that I employ standard natural deduction abbreviations of the inference rules, a strict distinction between premises (PREM) and hypotheses (HYP), and the hypothetical rule Conditional Proof (→I). All portions of hypothetical reasoning are clearly marked by vertical lines. Some of the predicates and premises (suppressed earlier) need some explanation. These explanations are provided right after the formalisation.

Key:

Kx: x is in domain K
Tx: S has to carry out task T regarding x
Rxy: S first has to carry out T regarding y in order to carry out T regarding x
Ax: S performs action A regarding x
Cx: S carries out T regarding x.

Example:

Kx: x is a dispute
Tx: S has to settle x
Rxy: S first has to settle y in order to settle x
Ax: S invokes a proposition to settle x
Cx: S settles x.

Failure Schema A

(1)	$\forall x(Ax{\rightarrow}(\exists yKy{\wedge}x{\neq}y))$	PREM
(2)	$\forall x(Kx{\rightarrow}Tx)$	PREM
(3)	$\forall x((Kx{\wedge}Tx){\rightarrow}\exists y(Ky{\wedge}Ty{\wedge}x{\neq}y)){\rightarrow}\neg\forall x(Kx{\wedge}Cx)$	PREM
(4)	$\forall x((Kx{\wedge}Tx){\rightarrow}Ax)$	HYP →I
(5)	$Ka{\wedge}Ta$	HYP →I
(6)	$(Ka{\wedge}Ta){\rightarrow}Aa$	4; $\forall$E
(7)	Aa	5, 6; →E
(8)	$Aa{\rightarrow}(\exists yKy{\wedge}a{\neq}y)$	1; $\forall$E
(9)	$\exists yKy{\wedge}a{\neq}y$	7, 8; →E
(10)	$Kb{\wedge}a{\neq}b$	HYP →I
(11)	Kb	10; $\wedge$E
(12)	$Kb{\rightarrow}Tb$	2; $\forall$E
(13)	Tb	11, 12; →E
(14)	$Kb{\wedge}Tb{\wedge}a{\neq}b$	10, 13; $\wedge$I
(15)	$\exists y(Ky{\wedge}Ty{\wedge}a{\neq}y)$	14; $\exists$I
(16)	$(Kb{\wedge}a{\neq}b){\rightarrow}\exists y(Ky{\wedge}Ty{\wedge}a{\neq}y)$	10-15; →I
(17)	$\exists y(Ky{\wedge}Ty{\wedge}a{\neq}y)$	9, 16; $\exists$E
(18)	$(Ka{\wedge}Ta){\rightarrow}\exists y(Ky{\wedge}Ty{\wedge}a{\neq}y)$	5-17; →I
(19)	$\forall x((Kx{\wedge}Tx){\rightarrow}\exists y(Ky{\wedge}Ty{\wedge}x{\neq}y))$	18; $\forall$I
(20)	$\neg\forall x(Kx{\wedge}Cx)$	19, 3; →E
(21)	$\forall x((Kx{\wedge}Tx){\rightarrow}Ax){\rightarrow}\neg\forall x(Kx{\wedge}Cx)$	4-20; →I

Failure Schema B

(1)	$\forall x(Ax \rightarrow \exists y(Ky \wedge Ty \wedge Rxy))$	PREM
(2)	$\forall x((Kx \wedge Tx) \rightarrow \exists y(Ky \wedge Ty \wedge Rxy)) \rightarrow \neg \exists x(Kx \wedge Cx)$	PREM
(3)	$\forall x((Kx \wedge Tx) \rightarrow Ax)$	HYP $\rightarrow$I
(4)	$Ka \wedge Ta$	HYP $\rightarrow$I
(5)	$(Ka \wedge Ta) \rightarrow Aa$	3; $\forall$E
(6)	Aa	4, 5; $\rightarrow$E
(7)	$Aa \rightarrow \exists y(Ky \wedge Ty \wedge Ray)$	1; $\forall$E
(8)	$\exists y(Ky \wedge Ty \wedge Ray)$	6, 7; $\rightarrow$E
(9)	$(Ka \wedge Ta) \rightarrow \exists y(Ky \wedge Ty \wedge Ray)$	4-8; $\rightarrow$I
(10)	$\forall x((Kx \wedge Tx) \rightarrow \exists y(Ky \wedge Ty \wedge Rxy))$	9; $\forall$I
(11)	$\neg \exists x(Kx \wedge Cx)$	2, 10; $\rightarrow$E
(12)	$\forall x((Kx \wedge Tx) \rightarrow Ax) \rightarrow \neg \exists x(Kx \wedge Cx)$	3-11; $\rightarrow$I

Premise (3) of Failure A and (2) of Failure B were suppressed in the semi-first-order schemas in Sect. 3.1 and require some further explanation. In terms of the latter schemas, they read:

- If S has to φ a new item in K for any item in K that S has to φ, then S will never φ all items in K.
- If S first has to φ a new item in K for any item in K that S has to φ, then S will never φ any item in K.

For example:

- If S has to settle a new dispute for any dispute that S has to settle, then S will never settle all disputes.
- If S first has to settle a new dispute for any dispute that S has to settle, then S will never settle any dispute.

According to this construction, the predicates 'T' and 'C' do not depend on each other. That is, if S has to carry out T, then she may or may not in fact carry out T.[7] Relatedly, 'T' does not carry modal or deontic connotations. At least, none of the inferences relies on such considerations. For example, they do not make use of the consideration that ought-implies-can (i.e. that if S has to carry out T, then S should be able to carry out T). According to the Failure Theory, 'S fails to carry out T regarding any/all K(s)' does not mean 'S lacks a certain ability', but rather 'S always has to accomplish a further task of the same sort in order to carry out T regarding any/all K(s), and so S will never carry out T regarding any/all K(s) in this sense'.

---

[7] Further research should show whether there are alternatives to this 'T'/'C' construction in premise (3) of Failure A and premise (2) of Failure B.

It might help to think of the notion of a 'potential infinity' in this context.[8] An actual infinity has infinitely many members. For example, the set of all natural numbers is an actual infinity. In contrast, a potential infinity has finitely many members, yet is *always* increasing in number. Regresses as understood within the Failure Theory are like the latter: the tasks that have to be carried out are always increasing in number, and in this sense one will never carry out any/all tasks.

The main difference between the two Failure Schemas lies in the predicate 'R'. 'R' cannot be expressed purely in terms of the predicate 'T', given that 'R' imposes an order on tasks (i.e. something that the tasks themselves do not have). Specifically, the term 'first' in 'S first has to settle dispute y in order to settle dispute x' (i.e. 'Rxy') indicates an instrumental order rather than a temporal order. It need not be the case that the problem of settling y must be solved earlier in time. What matters is the asymmetry between the problems: settling y is meant to be a precondition of settling x, and not the other way around. Formally: $\forall x \forall y (Rxy \rightarrow \neg Ryx)$.

Finally, Failure A's premises (1) and (3) explicitly assume that x and y are distinct items. First (3)'s antecedent would automatically be satisfied without this assumption (which is undesirable, because in that case the failure would follow at once). Second, in this schema we have no asymmetric relation between the tasks that can ensure that x = y. Yet, there still remains a problem about (3), as its antecedent does not say what it should say. It should say that there is always a *new* task of the same kind to be carried out, while in fact it merely says that for each task, there is a *distinct* task of the same kind to be carried out. To solve this, we could employ the '<'-assumptions from Sect. 2.4.

References

Aristotle. *Physics*. Trans. R. P. Hardie and R. K. Gaye 1930. Oxford: Clarendon.

Armstrong, D.M. 1974. Infinite regress arguments and the problem of universals. *Australasian Journal of Philosophy* 52: 191–201.

Armstrong, D.M. 1978. Realism and nominalism. In *Universals and scientific realism*. Vol. 1. Cambridge: CUP.

Black, O. 1996. Infinite regress arguments and infinite regresses. *Acta Analytica* 16: 95–124.

Bliss, R.L. 2013. Viciousness and the structure of reality. *Philosophical Studies* 166: 399–418.

Day, T.J. 1986. *Infinite regress arguments. Some metaphysical and epistemological problems*. Ph.D. Dissertation, Indiana University.

Day, T.J. 1987. Infinite regress arguments. *Philosophical Papers* 16: 155–164.

Dowden, B. 2013. The infinite. In *The Internet Encyclopedia of Philosophy*.

Feldman, R. 1993. *Reason and argument*, 2nd ed. 1999. Englewood Cliffs: Prentice-Hall.

Gamut, L.T.F. 1982. Introduction to logic. In *Logic, language and meaning*. Vol. 1. Trans. 1991. Chicago: UCP.

Gillett, C. 2003. Infinitism redux? A response to Klein. *Philosophy and Phenomenological Research* 66: 709–717.

---

[8] This notion traces back to at least Aristotle's *Physics*, book 3. Cf. Dowden (2013, Sect. 3.1).

Huggett, N. 2002. Zeno's paradoxes. In *Stanford Encyclopedia of Philosophy*, ed. E.N. Zalta.

Johnson, O. 1978. *Skepticism and cognitivism*. Berkeley: CUP.

Johnstone, Jr. H.W. 1996. The rejection of infinite postponement as a philosophical argument. *Journal of Speculative Philosophy* 10: 92–104.

Maurin, A.-S. 2007. Infinite regress: virtue or vice? In *Hommage à Wlodek*, eds. T. Rønnow-Rasmussen et al., 1–26. Lund University.

Maurin, A.-S. 2013. Infinite regress arguments. In *Johanssonian investigations*, eds. C. Svennerlind et al., 421–438. Heusenstamm: Ontos.

Nolt, J., D. Rohatyn and A. Varzi 1988. *Theory and problems of logic*. 2nd ed. 1998. New York: McGraw-Hill.

Passmore, J. 1961. *Philosophical reasoning*. 2nd ed. 1970. New York: Scribner's Sons.

Rankin, K.W. 1969. The duplicity of Plato's third man. *Mind* 78: 178–197.

Rescher, N. 2010. *Infinite regress. The theory and history of a prominent mode of philosophical argumentation*. New Brunswick: Transaction.

Rodriguez-Pereyra, G. 2002. *Resemblance nominalism. A solution to the problem of universals*. Oxford: Clarendon.

Rosenberg, J.F. 1978. *The practice of philosophy*. Englewood Cliffs: Prentice-Hall.

Ruben, D.-H. 1990. *Explaining explanation*. London: Routledge.

Russell, B. 1903. *The principles of mathematics*. 2nd ed. 1937. London: Allen & Unwin.

Sanford, D.H. 1984. Infinite regress arguments. In *Principles of philosophical reasoning*, ed. J.H. Fetzer, 93–117. Totowa: Rowman & Allanheld.

Schlesinger, G.N. 1983. *Metaphysics. Methods and problems*. Oxford: Blackwell.

Wieland, J.W. 2012. *And so on. Two theories of regress arguments in philosophy*. PhD dissertation, Ghent University.

Wieland, J.W. 2013a. Infinite regress arguments. *Acta Analytica* 28: 95–109.

Wieland, J.W. 2013b. Strong and weak regress arguments. *Logique & Analyse* 224: 439–461.

Chapter 4
Case Study: Carroll's Tortoise

As explained in Sect. 1.4, IRAs are often controversial regarding two issues: whether the suggested infinite regress is indeed generated, or, if it is agreed upon that it is generated, whether the suggested conclusion can indeed be drawn from the regress. In this chapter, I will present a case study to illustrate the first kind of controversy in some detail, namely a recent application of Lewis Carroll's famous case to the debate on rationality. (In the next chapter, I will provide an illustration of the second kind of controversy.) The main question throughout this chapter will be: is the suggested regress indeed generated? As we will see, this is a delicate issue and it is easy to be mistaken. At the end of the chapter, I will formulate a straightforward tool to check whether or not a given regress is generated.[1]

4.1 The Renewed Tortoise

Suppose I intend to publish a book, believe that publishing a book requires that I stay home this summer and write one, and yet refuse to intend to stay home and write one (while sticking to my initial intention and belief). I am irrational. To be rational is at least to have a consistent set of propositional attitudes. Yet, my three attitudes just listed are not clearly inconsistent. An extra story is needed that clarifies what sort of attitudes cannot be combined on pain of irrationality.

A renewed version of Carroll's Tortoise repeatedly shows up in this discussion. Following up on Brunero (2005), I ask what Carroll-style considerations actually prove in the rationality debate. Before explaining the renewed Tortoise, let me recall Carroll's initial puzzle.

---

[1] The material of this chapter derives from Wieland (2013).

J. W. Wieland, *Infinite Regress Arguments*, SpringerBriefs in Philosophy, DOI: 10.1007/978-3-319-06206-8_4, © The Author(s) 2014

Suppose you and I are debating the issue of whether Socrates is unsuited for a normal job, and you are making the following argument:

(a)   Socrates is a philosopher.
(b)   If Socrates is a philosopher, then he is unsuited for a normal job.
(z)   Socrates is unsuited for a normal job.

Now, suppose I am willing to accept (a) and (b), but not (z) just because I deny that (z) follows logically from (a) and (b) (so I am taking up the role of the Tortoise here). Furthermore, suppose that in order to demonstrate that (z) follows logically from those premises, you add the following, additional premise to the argument:

(c)   If (a) and (b) are true, then (z) is true.

Still, I am unsatisfied. This time I am willing to accept (a), (b) *and* (c), but not (z), just because I deny that (z) follows logically from (a), (b) and (c). To demonstrate that (z) follows logically given those premises, as the story goes, you add yet another premise to the argument:

(d) If (a), (b) and (c) are true, then (z) is true.
And so on.

Notoriously, Carroll himself did not draw any conclusion from this regress.[2] Thus commentators have attempted to identify what the Tortoise's lesson actually was. The general consensus here has been twofold.[3] On the one hand, the negative lesson is that if you add ever more premises to an argument (as above), then you will never demonstrate that its conclusion follows logically. On the other hand, the positive lesson is that only rules of inference will do the job, rather than premises of the form 'if premises such and such are true, then the conclusion is true'. In the example above, (z) follows logically from (a) and (b) thanks to the rule of inference Modus Ponens (rather than the premises (c), (d), etc.)[4]:

(R)   p, if p then q/q.

Moreover, once (R) is in place, it is thought, the regress disappears and all is solved. Unfortunately, even though the regress problem does disappear in Carroll's initial case, it can be shown that it does not disappear generally. For consider a slightly different piece of reasoning:

---

[2] Cf. "Here the narrator, having pressing business at the Bank, was obliged to leave the happy pair." (Carroll 1895, p. 280).

[3] For classic statements of this thought, cf. Ryle (1950) and Thomson (1960). Main subsequent commentaries include Stroud (1979) and Smiley (1995).

[4] Throughout this chapter, 'p' and 'q' are schematic letters to be replaced with full declarative sentences, and 'φ' and 'ψ', as before, with predicates which express actions.

(a)   I believe that Socrates is a philosopher.
(b)   I believe that [if Socrates is a philosopher, then he is unsuited for a normal job].
(z)   I ought to believe that Socrates is unsuited for a normal job.

This time, I do not reason on the basis of propositional contents, but on the basis of my attitudes towards such contents (beliefs, in this case). The idea is that if I have the beliefs described by (a) and (b), then it follows that I ought to adopt a further belief described by (z). I ought to adopt this belief in order to be rational and have a consistent set of attitudes. If I deny that I have to believe that Socrates is unsuited for a normal job, then I am irrational.

A similar Tortoise problem could be raised here. Suppose I accept (a) and (b), yet deny (z). How can it be shown that (z) follows from the premises I accept? Why do I have an obligation to have a certain belief given two other beliefs that I have? Unfortunately, the same solution that we used in Carroll's initial case is not immediately available. The reason is that, in this case, no classical rule of inference (such as Modus Ponens) will take us from (a) and (b) to (z). So, even if we assume that all classical rules of inference are in force, this renewed Tortoise problem remains unsolved. How is it the case that we ought to have certain propositional attitudes (beliefs, intentions, etc.) given certain other attitudes that we have?

Nevertheless, commentators have thought that Carroll-style considerations can still tell us something about this new problem. So the question is: what do Carroll-style considerations prove in the rationality debate? Against existing suggestions, I will argue that they highlight neither a premise/rule confusion, nor an internal/external confusion. Carroll's Tortoise shows something else here. Before explaining this (in Sect. 4.4), I will make explicit a number of assumptions in order to formulate more precisely the problem just sketched (Sect. 4.2), and introduce the main solutions to it (Sect. 4.3). As we will see, Carroll's Tortoise proves not that one of those solutions is false. Rather, it imposes a significant restriction on all of them.

4.2 Stage Setting

Here are five assumptions that will not be questioned in the rest of the chapter. They will allow me to formulate more precisely the problem sketched in Sect. 4.1.

First assumption: to keep the renewed Tortoise problem clearly distinct from the initial one, I will assume that all classical rules of inference are in force. This implies, as before, that we do not need to invoke the following premise in order to obtain 'q' from the premises 'p' and 'if p, then q':
(c)   If p and [if p, then q], then q.

As noted, however, in the following case (z) does not follow logically from (a) and (b) by classical rules:

Modus Ponens Consistency

(a)   S believes that p.
(b)   S believes that [if p, then q].
(z)   S ought to believe that q.

Nevertheless, (z) does follow classically from (a) and (b) plus the additional assumption that we ought to believe the consequences of the propositions we believe.[5] That is:

(c)   If S believes that p, and that [if p, then q], then S ought to believe that q.

Importantly, this does not mean that (c) *must* be part of one's reasoning, only that it *can* be part of it. As we will see in Sect. 4.3, there are solutions to the renewed Tortoise puzzle that do not make use of (c) (in fact, only one of the four solutions that I will discuss does so).

Second assumption: I will assume that all reasoning regarding one's attitudes concludes to obligations (rather something else). This assumption requires some further explanation. To begin with, all reasoning regarding one's attitudes will be called 'attitude reasoning'. The difference between ordinary and attitude reasoning is exemplified by the two argument patterns just described. Ordinary reasoning concerns what is or should be the case. Attitude reasoning, by contrast, is only indirectly concerned with this, and aims rather to cover all cases of reasoning where one is figuring out what propositional attitudes one should (or may) have. Attitude reasoning covers cases which concern figuring out what beliefs one should have, as well cases which concern figuring out what intentions one should have.[6] While Modus Ponens Consistency is a reasoning pattern of the former sort, the following is of the latter (further reasoning patterns of both varieties will be provided in due course):

Means/End Consistency

(a)   S intends to φ.
(b)   S believes that φ-ing requires S to ψ.
(z)   S ought to intend to ψ.

---

[5] It may be controversial to say that we ought to believe all the consequences of the propositions we believe (even if they are obvious). Less controversial would be to conclude from (a) and (b) to: S is permitted to believe that q and prohibited from disbelieving that q. For simplicity I will stick to the obligation formulation.

[6] Both beliefs and intentions are propositional attitudes and may have the same content. As Broome (2002, Sect. 2) points out, believing that p differs from intending that p in at least the following sense: the former attitude is one of taking it as true (or at least plausible) that p, whereas the latter attitude is one of making it true that p.

Here is the instance described at the very start of this chapter:

(a)   I intend to publish a book.
(b)   I believe that publishing a book requires that I stay home.
(z)   I ought to intend to stay home.

Not everyone agrees that practical reasoning like this concludes to *obligations* regarding one's attitudes, rather than directly to actions or intentions.[7] That is, one may alternatively conclude, in this case, to 'I intend to stay home' or even to 'I stay home' rather than to 'I ought to intend to stay home'. My main reason for the obligations variant is briefly the following. Failing to act or failing to adopt intentions may have all sorts of explanations (for example, one may be physically hindered to perform the given actions, or mentally unable to adopt the given intentions), yet denying one's obligations regarding one's attitudes is always irrational.

Third assumption: I assume, without explicitly stating it in each case, that all obligations in (z) could be 'wide-scope'; that I should adopt certain attitudes, that is, *if* I stick to my initial attitudes. Consequently, (z) can be resisted by retracting (a) or (b) in the course of one's reasoning. In the Modus Ponens Consistency example, for instance, if I do not wish to be obliged to believe that Socrates is unsuited for a normal job, then I could still retract my belief that Socrates is a philosopher or my belief that if he is a philosopher, he is unsuited for a normal job. Or, in the Means/End Consistency example, I should either accept my obligation to intend to stay home, or retract my initial intention to publish, or retract my belief that publishing requires that I stay home.[8]

Fourth assumption: I assume that the 'ought' in (z) is not an all-things-considered ought. It says that one ought to have certain attitudes given a specific set of other attitudes. It does not say that one ought to have certain intentions given sets of attitudes other than those considered, nor that one ought to have certain intentions generally (or all-things-considered). For example, if I intend to publish a book and believe that publishing requires that I bribe the editors, then I ought to bribe the editors. I ought to do this in order to be rational (i.e. to have a consistent set of attitudes), yet not in order to behave correctly in any other sense.

The problem of rationality that I am addressing in this chapter should be understood in terms of the assumptions so far: why should a given agent accept certain obligations regarding her attitudes (beliefs, intentions, etc.) given a certain set of other attitudes that she has? Why do I have the obligation to believe that Socrates is unsuited for a normal job given two of my other attitudes? Why do I have the obligation to intend to stay home given two of my other attitudes? Surely,

---

[7] For an overview of these options, cf. Streumer (2010).

[8] Of course, complex issues attach to this wide-scope phenomenon that cannot be discussed here. For an overview, cf. Way (2010). For example, is it really rational to refuse to intend to stay home by sticking to one's intention to publish and dropping one's belief that publishing requires one to stay home? Also, to what extent are we psychologically free, even if we are entitled, to retract intentions and beliefs in the course of our reasoning?

the triads (a), (b) and not-(z) look inconsistent, yet they are not inconsistent in any straightforward, classical sense. No contradiction follows, for example, from having the intention to publish plus the belief that publishing requires that one do certain things, while denying that one ought to intend to do that things.

This question is important. Clearly, there are correct and incorrect ways to reason on the basis of one's attitudes. We must accept some, yet not all sorts of obligations given the attitudes we happen to have. I should have the intention to stay home given my attitudes. Yet the following pattern should not, for instance, be expected to provide correct instances: S's evidence supports the belief that p/S ought to disbelieve that p.

Final assumption: following Brunero (2005), I assume that all forms of attitude reasoning require the same sort of solution. Thus, whatever kind of solution gets you to accept the obligation to believe in Modus Ponens Consistency will also get you to accept the obligation to intend in Means/End Consistency.

4.3 Four Solutions

Below, I will present a brief overview of four solutions to the problem just stated. Each will be explained in terms of the Means/End Consistency case (but the solutions can easily be generalized):

(a)   S intends to φ.
(b)   S believes that φ-ing requires S to ψ.
(z)   S ought to intend to ψ.

So what validates the step from (a) and (b) to (z)? The four upcoming solutions may be called External Premise, Internal Premise, External Rule and Internal Rule respectively. I will also identify a potential problem for each of the first three. They show why one should not easily be satisfied with them. That is, they will serve as an introduction to the next solution, yet play no further role in my discussion of Carroll's Tortoise later on.

Solution 1 (External Premise): S should accept (z) in the sense that S is logically committed to (z) if she sticks to (a), (b) and the extra premise (c):

(c)   If S intends to φ, and believes that φ-ing requires S to ψ, then S ought to intend to ψ.

A potential problem: why should S accept (c) as a premise in her reasoning if it is not among her beliefs?

Solution 2 (Internal Premise): S should accept (z) on the basis of (a), (b) and the extra premise (c*), which is now internalised, i.e. among S's beliefs:

(c*)   S believes that [if S intends to φ, and believes that φ-ing requires S to ψ, then S ought to intend to ψ].

A potential problem: S is no longer committed to (z), as it does not follow logically from (a), (b) and (c*). There is no classic rule which takes you from (a), (b) and (c*) to (z). On top of this, beliefs like (c*) are demanding, and presumably no-one besides philosophers ever entertained them. If so, only philosophers would have to deal with obligations regarding one's attitudes; which would be absurd.

Solution 3 (External Rule): S should accept (z) on the basis of (a), (b), and the extra rule of inference (R) which is to be applied to (a) and (b):

(R)   S intends to φ; S believes that φ-ing requires S to ψ/S ought to intend to ψ.

The difference with the previous solutions is that (R) is not a premise, either external or internal to S's beliefs, but a rule of inference. So (R) is to have the same status as Modus Ponens and other rules of inference, and says that S is logically committed to (z) on the basis of (a) and (b).

A potential problem: according to this solution, there has to be an extra rule of inference for each and every pattern of reasoning. Here is a list of some important rules (adapted from Way 2010):

- S believes that p; S believes that [if p, then q]/S ought to believe that q. (Modus Ponens Consistency)
- S intends to φ; S believes that φ-ing requires S to ψ/S ought to intend to ψ. (Means/End Consistency)
- S believes that p/S ought not to disbelieve that p. (Belief Consistency)
- S intends to φ; S believes that S cannot both φ and ψ/S ought not to intend to ψ. (Intention Consistency)
- S believes that she should φ/S ought to intend to φ. (Enkrasia)

This list is only partial and it is easy to think of many more cases by just playing around with beliefs, intentions, obligations, permissions, etc. But do we want to accept this explosion of inference rules? Surely we do not want to adopt the rule mentioned earlier: S believes that p/S ought to disbelieve that p. Hence, the challenge for External Rule proponents is to provide criteria that can tell us which rules are in force and which are not. Moreover, as we will see next, no such challenge exists for Internal Rule proponents.

Solution 4 (Internal Rule): S should accept (z) on the basis of (a), (b), (R) and S's pro-attitude towards (R).

The advantage of this solution over the previous one is that it blocks explosion. That is: not all possible rules of inference are supposed to be in force, but only those that S acknowledges. For S to have a pro-attitude towards (R) is for S to do more than just reason in accordance with (R) (which might just be coincidental or a mere regularity). Rather, it is to let one's reasoning be governed by (R). Or again: it is a desire on S's part to comply with (R) and apply it to (a) and (b). Pro-attitudes differ from beliefs at least in the following way: to believe (R) is to *regard* it as true that one ought to intend the means that one believes to be necessary to one's ends, while to have a desire to comply with (R) is to *want* it to be true that one's attitudes are governed by this rule.

Now recall our main problem: why should a given agent accept obligations to adopt certain attitudes given other attitudes that she has? At this point we have four options at our disposal: two which suggest a need for extra premises (one external, one internal), two which suggest a need for extra rules (one external, one internal). Moreover, two of the four solutions introduce additional attitudes, namely beliefs or pro-attitudes (i.e. the internal solutions), and two do not introduce such additional attitudes (i.e. the external solutions). The question is: can Carroll's Tortoise help us to see which solution is the right one?

Before turning to this question in Sect. 4.4, I want to be explicit concerning what the four solutions solve (if they succeed) and what they leave unaddressed. The solutions do form a candidate answer to the question of why we should accept certain obligations in order to be rational. However, they do *not* motivate why we should be rational in the first place, i.e. explain the value or importance of having consistent sets of attitudes.[9] Or again: they say something about why certain attitudes cannot be combined on pain of irrationality, but say nothing about why these governing elements should be accepted in the first place (i.e. why the premises should hold, or the rules be in force).

4.4 Three Hypotheses

As noted, commentators have attempted to identify what actually was the Tortoise's lesson in the rationality debate.[10] They generally agree that something might go wrong as soon as we introduce extra factors that are to account for the transition from our given attitudes to our obligations. The thought is: if our actual attitudes do not suffice to explain our obligation to hold a certain attitude, then why suppose that additional factors would be any more successful?

But the question is, of course, how this can be made precise. In the following I will demonstrate that this is a rather delicate issue, and that even Brunero's (2005) analysis, i.e. the most detailed to date, does not get the matter entirely right. To show this, let us consider the following three hypotheses. Carroll's Tortoise might demonstrate that:

(H1)   The solutions which introduce additional premises, rather than rules, are committed to a regress and hence fail.

(H2)   The internal solutions, which introduce additional attitudes, are committed to a regress and hence fail.

(H3)   The solutions which implicitly invoke additional obligations are committed to a regress and hence fail.

---

[9] For this vexed issue, cf. Broome (2005) and Kolodny (2005).

[10] Main pioneering applications of Carroll's Tortoise to practical reasoning are Blackburn (1995) and Schueler (1995). Below I will mainly focus on Dreier (2001) and Brunero (2005). Importantly, some of these discussions concern the internal versus external reasons for action debate, rather than the rationality debate. Nevertheless, as we will see below, the parallel is quite close: where the latter speak of 'obligations', the former speak of 'motivating reasons'.

Hence, (H1) says that Carroll's Tortoise refutes the solutions Internal and External Premise (but not Internal or External Rule). (H2) says that it refutes the solutions Internal Premise and Internal Rule (but not External Premise or External Rule). And (H3), as I will explain, says that it refutes none of the four solutions, but only specific versions of them. Furthermore, only the first two hypotheses (H1) and (H2) have been advanced in the literature. I will argue in the following, however, that only (H3) is correct.

4.4.1 Rule/Premise

A common and widespread interpretation of Carroll's Tortoise suggests (H1), i.e. that it draws our attention to a premise/rule confusion. Here is a clear formulation of this line:

> The lesson of Carroll's parable is that the refusal to accept a rule of inference cannot be compensated for by the addition of any number of premises—not even if one of these premises is an articulation of this rule of inference itself. (Schwartz 2010, p. 90)

As Brunero (2005, pp. 560–561) shows in response to Railton (1997, pp. 76–77), (H1) cannot be correct. Of course, in Carroll's initial case it is unhelpful to add extra premises (c), (d), etc. rather than apply Modus Ponens directly to (a) and (b) (i.e. in order to obtain (z)):

(a) p.
(b) If p, then q.
(c) If (a) and (b), then (z).
(d) If (a), (b) and (c), then (z).
...
(z) q.

Yet, once the classical rules such as Modus Ponens are in place, this same sort of solution does not work for the renewed puzzle. Of course, it is still possible to invoke rules of rationality, yet it is also possible to use premises instead. Consider my example:

(a) I intend to publish a book.
(b) I believe that publishing a book requires that I stay home.
(z) I ought to intend to stay home.

In this case, I can obtain the obligation (z) from my attitudes (a) and (b) in at least two ways. First option: I could invoke an extra premise, 'if S intends to φ, and believes that φ-ing requires her to ψ, then S ought to intend to ψ', and then derive (z) in a classical way (i.e. by the rules Universal Instantiation, Conjunction and Modus Ponens). Alternatively, I could invoke the rule of rationality, 'S intends to φ; S believes that φ-ing requires S to ψ/S ought to intend to ψ', and derive (z) directly on the basis of this rule. Both options are regress-free, and so Carroll style-considerations cannot be used to favour the rule-option here as well.

Here is Brunero's own illustration of this point:

(a)   S's evidence supports the belief that p.
(z)   S ought to believe that p.

Again, we can obtain (z) in two different ways: either we can refer to the extra premise that one ought to believe what one's evidence supports, or create a rule that validates the step from (a) to (z) (i.e. 'S's evidence supports the belief that p/S ought to believe that p', labelled by Brunero as '(H-E)').[11] Nothing about Carroll's Tortoise makes the first option implausible, and so it cannot be used to dismiss the extra premise solutions. As Brunero puts it:

> If the rules of logical inference are put into the premises, the pains of regress begin. But there is nothing wrong with putting (H-E) into the premises. It is precisely what the argument was missing! (2005, p. 560)

Generally the objection to (H1) is that in many cases there is nothing wrong with adding premises to an argument. Sometimes, and especially in those cases where the conclusion does not follow from the relevant premises on the basis of familiar rules of inference, it may well be helpful to add extra premises. Moreover, one might suggest that pieces of attitude reasoning are exactly among such cases (because there are no classical rules which take us from initial attitudes to obligations of having further attitudes).

4.4.2 Internal/External

The second hypothesis (H2) states that Carroll's Tortoise draws our attention to an internal/external confusion. The basic idea is that whatever is to govern our attitudes may be something additional, yet should not itself be among our attitudes, i.e. should be external to them. Indeed: if our actual attitudes do not suffice to obligate us to hold certain attitudes, then why suppose that additional attitudes (like beliefs and pro-attitudes, as Internal Premise and Internal Rule propose) could be of any help? Here is a clear formulation of this line by Wedgwood (commenting on Railton 1997):

---

[11] As Brunero himself notes, this case is a bit simplistic. Arguably, one has an obligation to believe what one's evidence supports regarding whether p only if one is going to have any opinion about whether p at all, one is not in bad (or even misleading) evidential circumstances, and there are no strong non-evidential reasons to disbelieve that p.

> Practical reasoning takes us from antecedent beliefs, intentions, and desires to forming a new intention about what to do. For practical reasoning to do this, we need to exercise a capacity (or manifest a disposition) for reasoning of the relevant sort. Exercising this capacity cannot consist in our having any further beliefs, intentions, or desires—on pain of a regress of the same sort as that into which the Tortoise led Achilles. (2005, p. 467)

Yet I will argue that (H2) cannot be right. First I will briefly explain the point in terms of Internal Premise, i.e. the view that we have obligations to accept certain attitudes given other attitudes that we have thanks to certain further beliefs. After that, I will explain why Carroll's Tortoise does not refute Internal Rule either. Consider again my example:

(a)   I intend to publish a book.
(b)   I believe that publishing a book requires that I stay home.
(z)   I ought to intend to stay home.

Now Internal Premise says that I have to accept (z) on the basis of (a), (b) and the following, further belief:

(c)   I believe that [if I intend to publish a book, and believe that this requires that I stay home, then I ought to intend to me to stay home].

Although it is not clear why (z) is supposed to follow from (a), (b) and (c) (noted in Sect. 4.3), it can be pointed out that Internal Premise has no regressive consequences. The following variant of Internal Premise, in contrast, would have the following consequence:

IR/1   S is obliged to adopt a new attitude x given other attitudes that she has only if S is obliged to adopt the additional belief that she ought to have x given those attitudes.

For if S is obliged to adopt an attitude (any attitude) only if she is obliged to adopt an additional belief, then she is equally obliged to adopt that additional belief only if she is obliged to adopt yet another additional belief, and so on. The regressive consequences of the upcoming cases will be spelled out in more detail, but the general idea seems clear.

Important here is that IR/1 is considerably stronger than Internal Premise itself, i.e. the view that merely says that S is obliged to adopt a new attitude x given other attitudes that she has only if S *has* the additional belief that she ought to have x given those attitudes. Thus, so long as IR/1 forms no essential part of Internal Premise, the latter is regress-free and does not fail for this reason. Moreover, this invalidates hypothesis (H2), which locates a problem in additional attitudes generally.

I will argue that this same problem (i.e. the problem that the internal solutions may well be regress-free) afflicts the defences of (H2) by respectively Dreier (2001) and Brunero (2005). Their main target is Internal Rule, rather than Internal Premise, just discussed. Let us first consider Dreier's case:

(a)   Ann desires to go a good law school.
(b)   Ann believes that by taking a prep course she will go a good law school.
(z)   Ann has a motivating reason to take a prep course. [by (M/E) plus Ann's desire to comply with (M/E)]
(M/E)   S desires to φ; S believes that by ψ-ing S will φ/S has a motivating reason to ψ.

Internal Rule would say that Ann should accept that she has a motivation to take a prep course on the basis of (a), (b) and her desire to comply with the rule (M/E). The role of such desires, as I see it, is to authorize a rule of rationality (see Sect. 4.3). If one has a desire, in this case, to comply with (M/E) then one wants to organize one's attitudes in accordance with it. Now, Dreier disagrees with Internal Rule's diagnosis, and holds that Ann can still refuse to be motivated to take a prep course:

> Were she to desire to comply with (M/E), would she then be motivated to take the LSAT prep course? By hypothesis, Ann suffers from this failure of practical reason: she fails to be motivated by the acknowledged means to her desired ends. So adding a desire (complying with (M/E)) does not in her bring about the motivation to perform an acknowledged means to her end of doing well in the LSAT. (2001, p. 39)

Thus, Dreier reasons (and Jollimore (2005, pp. 293–295) concurs), if Ann's desire to go a good law school does not motivate her to take all the steps that she believes are required or appropriate for this end (such as taking a prep course), then we should not expect Ann's additional desire to comply with (M/E) to help her in this.

Yet I believe that this analysis is incorrect. Dreier says that Ann "fails to be motivated by the acknowledged means to her desired ends." Yet, the scenario does not say, and Internal Rule does not require, that Ann's additional desire to comply with (M/E) be among the acknowledged means to her desired ends, and hence it does not follow that she fails to be motivated by her additional desire to comply with (M/E). Surely, the following assumption would be troublesome:

IR/2   Ann is motivated to φ only if she is motivated to comply with (M/E) (or a rule quite like this).

According to this, Ann is motivated to take a prep course only if she is motivated to comply with (M/E). Similarly, by IR/2 she is motivated to comply with (M/E) only if is motivated to comply with a rule quite like (M/E):

(M/E*)   S desires to φ; S believes that by ψ-ing S will φ; S desires to comply with (M/E)/S has a motivating reason to ψ.

It is easy to see how the regress would continue. However, nothing about Internal Rule is committed to IR/2. Internal Rule merely says that Ann is motivated to φ only if she has a desire to comply with (M/E). She need not be motivated to comply with (M/E) in addition to this.[12] Here is Brunero's clarification of Dreier's reasoning:

> The Tortoise's acceptance of Modus Ponens as a premise was futile because the Tortoise refused to apply Modus Ponens to his premises. (Likewise, Ann's attainment of a desire to comply with (M/E) is futile because Ann, by hypothesis, refuses to apply (M/E) to her desires and beliefs.) (2005, p. 562)

Unfortunately, however, this does not help, as the analogy breaks down. It is true that the Tortoise refuses to apply Modus Ponens to her premises. Yet the Tortoise has no desire to comply with Modus Ponens, and (unless we have more information about the situation) might well refuse to apply it. Ann, by contrast, does have a desire to comply with (M/E). Within the Internal Rule framework, Ann cannot deny that she has a motivating reason to take the prep course, given the other attitudes she has.

Next I will show that a similar problem returns for Brunero's own motivation for (H2). Brunero's analysis differs from Dreier's in two main respects. First, where Dreier speaks of 'motivating reasons' (i.e. for action), Brunero speaks of 'obligations' (i.e. to adopt certain propositional attitudes). Second, Brunero takes his reasoning to apply to *all* rules of rationality, and not just to (M/E). Dreier argues that the (M/E) rule must have a special status. Specifically, his view is that instrumental rationality is fundamental, as we cannot have pro-attitudes towards (M/E) while we can have such attitudes towards other such rules. In response, Brunero shows that the same kind of reasoning can be used in favour of any rule, and not just regarding (M/E). To explain this, let us consider my Means/End Consistency case again (yet I could have taken any other case of attitude reasoning as well):

(a)   I intend to publish a book.
(b)   I believe that publishing a book requires that I stay home.
(z)   I ought to intend to stay home.
(R)   S intends to φ; S believes that φ-ing requires S to ψ/S ought to intend to ψ.

By Internal Rule, I am not obliged to apply (R) to my given attitudes (a) and (b) unless I have the additional desire to comply with (R). If that is so, we seem to have the following, new situation:

(a)   I intend to publish a book.
(b)   I believe that publishing a book requires that I stay home.
(c)   I desire to comply with (R).
(z)   I ought to intend to stay home.
(R*) S intends to φ; S believes that φ-ing requires S to ψ; S desires to comply with (R)/S ought to intend to ψ.

---

[12] Note that, on Dreier's account (2001, p. 35), to be motivated is not just to have a desire, but also to have a belief about how to satisfy that desire.

Brunero's reasoning is this (2005, p. 563ff): If I am not obliged to apply (R) to my given attitudes (a) and (b) unless I have the additional desire to comply with (R), then by parity of reason it is similarly the case that I am not obliged to apply (R*) to my new set of attitudes consisting of (a), (b) and (c) unless I have the additional desire to comply with (R*). It is easy to see how this reasoning can be continued. Furthermore, if ever more additional desires must be supplied in order to bridge the gap between my attitudes (a) and (b) and obligation (z), then we can never cross this gap, so to speak. In brief: the additional attitudes provided by Internal Rule are idle.

This reasoning fails basically in the same way that Dreier's initial argument failed. The problem is that (R*) and further rules, plus desires to comply with them, are irrelevant. It is true that, by Internal Rule, I am obliged to apply (R*) to (a), (b) and (c) only if I have the additional desire to comply with (R*). However, this does not matter so long as my obligation to apply (R) does not depend on any obligation to apply (R*). Nothing about Internal Rule is committed to this. Internal Rule says that I am obliged to apply (R) only if I have a desire to comply with it, and does not also require that I apply a new rule, i.e. (R*), to my new, expanded set of attitudes.

Still, even if it does not lend support for (H2), there is something about Brunero's analysis that does lead in the right direction.

4.4.3 Obligations

Here is, I take it, the crucial passage from Brunero's article:

> In my view [...] the potential for a Carroll-style regress just shows us that since instru-
> mental rationality involves a higher-order commitment to combine our willing an end with
> our taking the necessary means, it therefore cannot, on pain of regress, itself be added as a
> conjunct to one of the elements to be combined. (2005, p. 563)

As just argued, this fails if it is spelled out along the lines of (H2). Carroll's Tortoise does not show that further attitudes cannot be among the attitudes to be combined in order to result in an obligation. There is, however, one crucial notion in this passage, i.e. that of a 'higher-order commitment', that I will use to defend (H3) instead (i.e. the view that the solutions which implicitly invoke additional obligations are committed to a regress and hence fail).

In Sect. 4.4.2, we have seen that Internal Rule generates no regress. Still, it can be shown that the following does:

IR/3   I am obliged to apply a rule x, any rule, to my attitudes only if I am obliged to apply another rule y to a bigger set of attitudes containing, in addition, the attitude that I desire to comply with x.

This generates a regress of obligations. By IR/3, I am obliged to apply (R) to my attitudes only if I have the following higher-order obligations:

- to apply (R*) to a bigger set of attitudes containing, in addition, the attitude that I desire to comply with (R);
- to apply (R**) to a bigger set of attitudes containing, in addition, the attitude that I desire to comply with (R*);

etc.

where

(R)   S intends to φ; S believes that φ-ing requires S to ψ/S ought to intend to ψ;

(R*)   S intends to φ; S believes that φ-ing requires S to ψ; S has a pro-attitude towards (R)/S ought to intend to ψ;

(R**)   S intends to φ; S believes that φ-ing requires S to ψ; S has a pro-attitude towards (R); S has a pro-attitude towards (R*)/S ought to intend to ψ;

etc.

These obligations are called 'higher-order' because they oblige one to have pro-attitudes towards higher-order rules, i.e. rules which are about other rules. They are not intended to be reflective: the agent that has them need not be aware of them. Relatedly, these higher-order obligations are not meant to be problematic because they would have to figure explicitly in our reasoning. Rather, they are problematic because IR/3 establishes that I am obliged to apply (R) only if I am obliged to apply an endless number of rules to an endless number of sets of my attitudes. This implies that I will never be in the position required to be obliged to apply (R). Hence, if IR/3 were in place, then I would not be obliged to apply (R), or indeed any other rule. But clearly I am obliged to apply a number of rules, so we should look for an alternative theory that is not committed to IR/3 (and does not invoke higher-order obligations).

Generally the idea is that Carroll-style considerations show that our obligations to apply rules to our attitudes, and indeed our obligations to adopt certain attitudes given other attitudes that we happen to have, should not depend on any further obligations of the same sort (on pain of infinite regress). This, indeed, is (H3).

One might wonder whether this has anything to do with Carroll's initial puzzle. But it does. And the parallel is actually quite close. For the most direct lesson of Carroll's initial puzzle is *not* that we should employ rules of inference in order to draw conclusions from premise sets, but rather that our obligations to accept conclusions given certain premise sets should not depend on further obligations of the same sort, i.e. of obligations to accept conclusions given bigger premise sets (on pain of infinite regress). More precisely, it shows that the following line creates trouble:

IR/4   The Tortoise is obliged to accept a conclusion x given a set of premises y, any set, only if it is obliged to accept x given a bigger set containing, in addition, the premise 'if the members of y are true, then x is true'.

Again, this generates a regress of obligations, and in the same way as above it establishes that the Tortoise is not obliged to accept whatever conclusion. As the Tortoise is clearly obliged to accept plenty of conclusions given the premises it subscribes to, IR/4 should be abandoned. (Only at this point does the positive lesson about rules become clear: we must accept conclusions from a premise set thanks to rules of inference, i.e. rather than premises of the form 'if the premises are true, then the conclusion is true'.)

The case can be generalised. As I will show below in Sect. 4.4, any view that subscribes to a line like IR/3 or IR/4 runs into regress problems. If so, this poses a restriction on any solution to our problem of how to generate obligations from our attitudes (namely, as I will explain, that it should entail no line like IR/3 or IR/4).

Nevertheless, this need not rule out any of the four solutions listed in Sect. 4.3. This means that, so far as Carroll's Tortoise is concerned, obligations can be generated from our attitudes via premises as well as via rules, supplemented with or without additional attitudes. The Tortoise rules out only versions of them that subscribe to a line like IR/3 or IR/4. The latter are, however, additional commitments and need not form part of the solutions themselves. Thus, as far as the Tortoise is concerned (and this might be surprising for Dreier, Brunero and other commentators in this debate) Internal Rule can be a perfectly viable solution.

4.4.4 General Diagnosis

So far I have argued that Carroll's Tortoise provides support for the hypothesis (H3), rather than for (H1) or (H2). Before concluding, I will demonstrate that my argument in the foregoing relies on a general fact about IRAs. This is not meant to lend further support to my argument, but merely to supplement it with a logical rationale.

That rationale is as follows. In all generality, regresses are generated by all instances of:

IR For any item x of a certain type, S φ-s x only if

(i)   there is a new item y of that same type, and
(ii)   S φ-s y.

In other words: for any task or feature X of a certain sort, you carry out or possess X only if there is another, distinct task or feature Y of that very same sort and you carry out or possess Y. Here is an overview of the instances that we have seen in this chapter (the last will be introduced below):

IR	Domain	$\varphi\, x$
1	Attitudes	Being obliged to adopt x
2	Actions	Being motivated to perform x
3	Rules	Being obliged to comply with x
4	Sets of premises	Being obliged to accept a conclusion given x
5	Propositions	Being justified in believing x

There are at least four important aspects of IR and if one of them is missing, no regress will ensue[13]:

- Its universal quantification (so that the principle applies to all new items of the given type);
- the fact that the items are of the same type (so that the new items fall within the scope of the principle);
- clause (i); and
- clause (ii).

Compare the familiar regress of reasons, which can be generated by another instance of IR:

IR/5   For all propositions x, S is justified in believing x only if (i) there is another proposition y (namely, a reason for x), and (ii) S is justified in believing y.

Suppose S is justified in believing a proposition p_1. By (i), S has a reason, p_2, for p_1. By (ii), S is justified in believing p_2. By (i) again, S has a reason, p_3, for p_2. And so on. This regress would not ensue if IR/5 were not universally quantified, if reasons were no propositions as well, if clause (i) were to fail, or lastly if clause (ii) were to fail.

The IRAs by Dreier and Brunero fail only because clause (ii) does not hold (even though all the rest is in place). In Dreier's case, Ann is motivated to perform an action only if she has a desire to comply with (M/E). Yet nothing follows, as we have seen, because it is not also required that Ann is motivated to comply with (M/E). In Brunero's case, S is obliged to comply with (R) only if S has a desire to comply with (R). Again, nothing follows so long as S need not be obliged to comply with another rule (R*) in addition to this.

4.5 Concluding Remark

Should I intend to stay home this summer? Why should I adopt certain attitudes given certain other attitudes that I have? Many have believed that Carroll's Tortoise has something important to say about this problem. I agree. Yet, the

---

[13] For further details, see Failure Schema B in Chap. 3. As we have seen in that chapter, a slightly different story holds for Failure Schema A (where clause (ii) does not apply, but a different kind of premise does).

importance of the Tortoise does not lie where commentators usually think it does. As I argued in this chapter, the Tortoise does not demonstrate that no extra premises (rather than rules) should be introduced in our reasoning, nor that whatever is to govern our attitudes (premises or rules) should remain external to our attitudes. Rather, it shows that no solution to this problem should entail an instance of IR, and that we should let our obligations to adopt certain attitudes depend on additional obligations to adopt further attitudes.

References

Blackburn, S. 1995. Practical tortoise raising. *Mind* 104: 695–711.

Broome, J. 2002. Practical reasoning. In *Reason and nature. Essays in the theory of rationality*, eds. J. Bermùdez and A. Millar, 85–111. Oxford: OUP.

Broome, J. 2005. Does rationality give us reasons? *Philosophical Issues* 15: 321–337.

Brunero, J. 2005. Instrumental rationality and Carroll's tortoise. *Ethical Theory and Moral Practice* 8: 557–569.

Carroll, L. 1895. What the tortoise said to Achilles. *Mind* 4: 278–280.

Dreier, J. 2001. Humean doubts about categorical imperatives. In *Varieties of practical reasoning*, ed. E. Millgram, 27–47. Cambridge: MIT.

Jollimore, T. 2005. Why is instrumental rationality rational? *Canadian Journal of Philosophy* 35: 289–307.

Kolodny, N. 2005. Why be rational? *Mind* 114: 509–563.

Railton, P. 1997. On the hypothetical and non-hypothetical in reasoning about belief and action. In *Ethics and practical reason*, eds. G. Cullity et al., 53–79. Oxford: Clarendon.

Ryle, G. 1950. 'If', 'so', and 'because'. In *Philosophical analysis*, ed. M. Black, 302–318. Ithaca: CUP.

Schueler, G.F. 1995. Why 'oughts' are not facts (or what the tortoise and Achilles taught Mrs. Ganderhoot and me about practical reason). *Mind* 104: 713–723.

Schwartz, J. 2010. Do hypothetical imperatives require categorical imperatives? *European Journal of Philosophy* 18: 84–107.

Smiley, T. 1995. A tale of two tortoises. *Mind* 104: 725–736.

Streumer, B. 2010. Practical reasoning. In *A companion to the philosophy of action*, eds. T. O'Connor and C. Sandis, 244–251. Malden: Wiley.

Stroud, B. 1979. Inference, belief, and understanding. *Mind* 88: 179–196.

Thomson, J.F. 1960. What Achilles should have said to the tortoise. *Ratio* 3: 95–105.

Way, J. 2010. The normativity of rationality. *Philosophy Compass* 5: 1057–1068.

Wedgwood, R. 2005. Railton on normativity. *Philosophical Studies* 126: 463–479.

Wieland, J.W. 2013. What Carroll's tortoise actually proves. *Ethical Theory and Moral Practice* 16: 983–997.

Chapter 5
Case Study: Access and the Shirker Problem

As explained in Sect. 1.4, IRAs are often controversial regarding two issues: whether the suggested infinite regress is indeed generated, and whether the suggested conclusion can indeed be drawn from the regress. In Chap. 2, we have seen an illustration of the first kind of debate. In the present chapter, we will turn to the second kind. For even if it is clear and agreed upon that a theory does entail a regress, then it is often controversial what conclusion should be drawn from it. What do regresses actually prove? There will be two, related questions throughout this chapter: first, what possible conclusions can regresses have? And second: how can those conclusions be defended on the basis of a regress, and how might those conclusions be resisted? Again, these are delicate issues, and we will employ the theories from Chaps. 2 and 3 to clarify and to resolve them.

This time, the case study will be a regress concerning the so-called Access principle (from ethics). This principle places an epistemic restriction on our obligations: it states that we have an obligation only if we can know that we have that obligation. Despite their initial plausibility, epistemic restrictions like Access fall prey to the 'Shirker Problem', namely the problem that shirkers could evade their obligations by evading certain epistemic circumstances. To block this problem, it has been suggested that shirkers have the obligation to learn their obligations. This solution yields a regress, yet it is controversial what the moral of the regress actually is. In this chapter, we will go into this issue in some depth.[1]

5.1 Access

We have plenty of obligations to do and to refrain from doing things. The question is whether our obligations should be knowable, or whether there exist such things as 'unknowable obligations'. For example, should one be in the position to know that one has the obligation to refrain from buying certain things (because they are

---

[1] The material of this chapter derives from Wieland (2014).

J. W. Wieland, *Infinite Regress Arguments*, SpringerBriefs in Philosophy, DOI: 10.1007/978-3-319-06206-8_5, © The Author(s) 2014

the product of child labour) in order to have that obligation in the first place? Those who think that this is required subscribe to the following principle:

Access. For all obligations x, an agent S has x only if S *can know* she has x.

In other words, this principle requires that my obligations are epistemically accessible to me. As there exist various readings of this principle, I will briefly specify how I understand it. First: Access is supposed to apply widely and range over obligations of all kinds. It applies to obligations regarding actions as well as omissions, to moral, epistemic, and other kinds of obligations, as well as to obligations to be in certain epistemic circumstances (as we will see).

How accessible are my obligations supposed to be? On the one hand, Access should be contrasted with the principle that one has an obligation only if one *actually* knows that one has it. I might be very ignorant (say because I lack curiosity and never ask around), yet this does not immediately excuse me from my obligations insofar as Access is concerned. On the other hand, Access should also be contrasted with the principle that one has an obligation only if it is *somehow* possible to know that one has it. Suppose I could possibly know I have a certain obligation, but only by carrying out expensive and time-consuming research. In that case, I will not say that the obligation is accessible to me.

Hence Access will roughly be taken as holding the following: that I have an obligation only if I can know I have it *without exerting much effort* (i.e. by taking a few inquisitive steps, such as asking a question, making an inference, paying attention, or carrying out a test).[2] In some cases, for sure, it is not entirely clear whether the knowledge of an obligation is easily accessible. In such a case, it follows from Access that it is indeterminate whether one has the obligation.

Relatedly, I will assume that Access is (roughly) synchronic: I have an obligation at a certain time t only if I can know I have it without exerting much effort at or just before t. I do not have an obligation if I can come to know it at some point in the far future (or in the far past for that matter), but not at or just before t itself. Clearly, this take on Access immediately raises questions of whether one is culpably ignorant of one's obligation, and how this might affect whether or not one has the obligation (an issue to which we will turn soon).

Finally, one might suggest that Access is false by referring to a distinction between obligations and *blameworthiness*. That is, if I cannot know I have an obligation, then although I may still have that obligation I am not to be blamed if I violate it. Yet even on this view, there is still an Access principle for blameworthiness[3]:

---

[2] Arguably, in certain cases (namely, when the moral stakes are high) one should take more than a few inquisitive steps (cf. Guerrero 2007).

[3] Cf. Sider (1995, p. 274). My formulation is slightly different, yet the idea is the same. This distinction traces back to at least Moore (1912, p. 100), cf. Graham (2010), Driver (2012) for recent defenses.

Access for Blame. For all obligations x, S is blameworthy for violating x only if S can know she has x.

In the following I will conceive of Access in terms of obligations, yet everything I will say carries over to this different principle (see Sect. 5.5 for more on this). As we can see, Access involves a major philosophical topic: when and in what sense our restricted epistemic viewpoint justifies or excuses us for certain things we have done, but might not have done had we had more information.[4] Despite their initial plausibility, however, epistemic restrictions like Access involve a regress problem. The aim of this chapter is two-fold. First, I employ logical tools to clarify and spell out this problem as it arose in the debate between Sorensen (1995) and Sider (1995). Second, on the basis of my analyses I propose solutions to the regress problem and conclude that the regress need not pose a problem for Access (that is, as we will see, so long as certain specific obligations are in place).

5.2 The Loophole

Sorensen (1995, pp. 254–256) rejects Access because it 'dumbs down ethics'.[5] To see this, consider the following consequence of Access. If I cannot know I have an obligation, then it is not the case that I have that obligation. Moreover, if I eliminate the possibility of my coming to know that I have an obligation, then I eliminate my (potential) obligation.

Sorensen's example is the following: I am obliged to donate some of my inheritance to charity only if I can know that I am obliged to donate. If I cannot know that I am obliged to donate, therefore, I am not obliged to donate. Now suppose that there is a will, and that this will is my only way to find out whether I have any obligations regarding the money. Furthermore, if I burn the will before reading it, then, assuming the will was my only access, I eliminate my obligation to donate. The synchronic character of Access is important in this case: even though the obligation to donate was accessibility to me in the past, it is no longer accessible to me *now*.

Let us define 'shirkers' as people who evade their obligations by making or keeping them unknowable. This definition includes two kinds of agents. The first is an agent that *makes* obligations unknowable by worsening her epistemic position (deliberately or otherwise). Sorensen's case is an example of this first kind. The second is an agent that *keeps* obligations unknowable by not improving her

---

[4] Main recent contributions in this debate on the interplay between obligations and/or responsibility on the one hand and the epistemic condition on the other (or: the extent to which obligations and/or responsibility are 'subjective') include Zimmerman (2008), Sher (2009), Smith (2010).

[5] Among a few other considerations. In this chapter, I focus exclusively on this main problem.

epistemic position (again, deliberately or otherwise).[6] My next case will be an example of this second kind. Now, if Access is true, both kinds of agents have fewer obligations than other people have. This is an unwelcome consequence of Access. Furthermore, the question is whether Access has to go given this problem. Call this problem for Access proponents the 'Shirker Problem'.

Here is the example I will use in this chapter. Suppose I would like to buy new shoes. Unbeknownst to me, the shoes I am interested in are made by a child in Indonesia. Had I watched the news the week before, I would have known that they were made under suspect circumstances. Yet I did not watch the news, and in the following days the media has lost interest in this issue. Assuming that I am no longer in the position to know about the child labour, it follows from Access (as defined in Sect. 5.1) that I do not have the obligation to refrain from buying the shoes. (There may still be a possible way to come to know that the shoes are made by the child, yet this knowledge is not easily accessible to me.)

In this debate, Sider (1995, p. 278) proposes the following solution to save Access, which I will call 'Block':

> *Block.* I should refrain from making my obligation to refrain from buying the shoes unknowable.

Is this a good solution? According to Sorensen, it is not, because Block together with Access generates a *regress*. To see this, I should say that Block concerns a new obligation. Now, according to Access we have this obligation only if it is knowable. Shirkers will try to make it the case, of course, that Block is unknowable to them so they will not have to do what Block says. To solve this new problem, we could introduce a new obligation:

> *SuperBlock.* I have the obligation to refrain from making Block unknowable.

Yet here the same story applies: shirkers will now try make it the case that SuperBlock is unknowable to them, which means that, according to Access, they will not have to do what SuperBlock says. So again, we are left with a new problem; and so on into the regress.

As Sorensen states the problem:

> But now a higher order loophole opens. If I keep ignorant of whether there is an obligation to ascertain my obligations, I can use the access principle to evade those epistemic obligations even if they exist. To close this meta-loophole, the defender of Access must invoke a yet higher order principle to the effect that we have an obligation to learn whether we have an obligation to learn our obligations (1995, p. 255).

Sorensen actually refers to a stronger version of Block: S has the obligation to *learn* her obligations. I will carefully distinguish these two versions of Block in Sect. 5.3, and will also have more to say about the scope of Block and SuperBlock.

---

[6] See Sect. 5.3 for more on this distinction.

At any rate, given Sorensen's analysis, we can understand the situation with the shoes as follows:

Scenario (1)
Obligation A: I ought to refrain from buying the shoes. As a shirker, I make it impossible to know whether I have A. By Access, therefore, I do not have A.

Scenario (2)
Obligation B: I ought to refrain from making it impossible to know whether I have A. As a shirker, I make it impossible to know whether I have B. By Access, therefore, I do not have B.
etc.

Conclusion: on the basis of the loophole in Access, shirkers can evade all their obligations, even if Block is in place.

Sider, responding to Sorenson, disagrees. According to him, the situation is rather the following:

Scenario (1)
Obligation A: I ought to refrain from buying the shoes. As a shirker, I make it impossible to know whether I have A. Where do I go wrong? I violate obligation B: I ought to refrain from making it impossible to know whether I have A.

Scenario (2)
As a shirker, I make it impossible to know whether I have B (see above). Where do I go wrong? I violate obligation C: I ought to refrain from making it impossible to know whether I have B.
etc.

Sider agrees that, in scenario 1, *after* I have made A unknowable, I no longer have A. He also agrees that, in scenario 2, *after* I have made B unknowable, I no longer have B. And so on. Yet this does not affect the fact that there is *always* an explanation of where the shirker goes wrong. Sider concludes that none of these scenarios puts any pressure on Access, or on Block for that matter. As he writes:

> At best, the regress consists of an infinite sequence of cases, none of which refutes Access (1995, p. 279).

The controversy between Sorensen and Sider can be summarized in four steps. First, Sorensen addresses the Shirker Problem about Access. Second, Sider defends Access by Block. In response, Sorensen raises the regress problem with Block. Last, Sider accepts the regress, yet denies that this is a problem for Block or Access.

As is clear, the debate has reached an impasse. Both philosophers agree there is a regress, yet disagree about its conclusion. Sorensen concludes that Block is unable to save Access and, as we saw, rejects the latter because it can be abused. Then Sider does acknowledge the regress, yet denies that it is vicious, i.e. that it forms a problem for Access. Now the question is: *who is right?* Does the regress make Access a bad restriction on our obligations or not? Or again: what exactly does the regress show?

The main goal of the chapter is *to get at the heart of the matter*. In Sects. 5.3 and 5.4, I will employ the tools from Chaps. 2 and 3 to clarify and resolve this intricate controversy. As we will see, these tools focus on different parts of the controversy and introduce some new elements to the debate.

5.3 Paradox Analysis

According to the Paradox Theory, IRAs demonstrate that certain propositions are false because they have regressive consequences that conflict with independent considerations (see Chap. 2). Let us consider the IRA against Access as reconstructed along the lines of this paradox idea:

Access (Paradox A instance)

(1) For all obligations x, I have x only if I can know I have x.
(2) For all obligations x, I can know I have x only if I have the obligation y to refrain from making x unknowable/to learn x.
(3) I have at least one obligation.
(4) I have an infinity of obligations, and can know all of them. [from 1–3]
(5) (4) is false: I do not have an infinity of obligations and/or cannot know all of them.
(C) (1) is false: It is not the case that for all obligations x, I have x only if I can know I have x. [from 1–5]

The goal of this argument is to refute Access: line (1). The line of reasoning is as follows: (1) is a hypothesis taken into consideration only for the sake of refuting it in the conclusion. (2) and (3) are premises accepted by the proponent of (1). (4) follows from the foregoing. Furthermore, it is shown in line (5) that (4) is false (because of independent evidence), so that the hypothesis we started with is committed to a contradiction and so is rejected in (C) (by Reductio Ad Absurdum).

Premise (2) requires some explanation. It is the generalized idea behind Block, and comprises two different principles[7]:

Weak Block. S has the obligation to refrain from making her obligations unknowable.

Strong Block. S has the obligation to make it possible to know, or even to learn, her obligations.

As the names indicate, Strong Block is stronger than Weak Block. Strong Block states that we should improve our epistemic perspective regarding our obligations, i.e. that we make it possible to them. Strictly speaking, making something knowable is not (always) the same as learning it. For example, if a certain secret report is unknowable unless I pay a lot of money for it, then I could refrain from

---

[7] Do Weak Block and Strong Block fall within their own scope? For the moment, we do not need to settle this. See Sect. 5.4.

keeping it unknowable by paying the money. But I only learn the information if I actually read the report. In contrast, Weak Block merely states that we are prohibited from weakening our epistemic perspective regarding our obligations. In the latter case, we are allowed to do nothing about, and so not improve, our epistemic perspective (i.e. we are allowed to keep certain obligations unknowable, if we are presently not in the position to know them). Simply put: Strong Block does not merely require that we not worsen our knowledge about our obligations, but also that we actively gather more information. The distinction is important, and I will have more to say about it, and the details of Strong Block and Weak Block, later when I return to the Shirker Problem.

Strictly speaking, given that one's knowledge of a certain obligation might come for free, premise (2) taken on itself makes little sense, and should be seen in conjunction with (1): if Access is in place, then I can know I have obligation A only if I have the obligation B to refrain from making A unknowable. Furthermore (2) seems accepted by Sider (at least in Weak Block version), given how he sets up his scenarios. For example, if I make it impossible to know whether I have A, I violate an obligation I *already* have, namely obligation B (i.e. to refrain from making it impossible to know whether I have A). By parity of reasoning, I *already* have an infinity of such obligations.

On the basis of this reconstruction it is clear how the regress could form a problem for Access: it is a direct refutation of it. Furthermore, on the basis of the reconstruction it can easily be seen what Sider (whose position is that the regress poses no problem for Access) could do to resist the argument. Namely, Sider could deny premise (5) and accept that one has an infinity of obligations and can know all of them. Is this plausible?

In my view, (5) is very controversial. For obligations to know and possibilities of knowing often come for free: why should there be any limit to the obligations I have? And why should there be any limit to the things I can possibly know? To answer these questions, we should carefully distinguish four lists of consequences generated by (1–3): those generated by Weak Block as opposed to Strong Block, and those that concern obligations to know as opposed to possibilities of knowing. First consider the regress of Weak Block obligations:

- I should refrain from buying the shoes;
- I should refrain from making the above obligation unknowable;
- I should refrain from making the above obligation unknowable;
etc.

This first list is not absurd at all. For according to these obligations, I do not have to do a *single* thing. I have merely to refrain from doing things altogether. But that is easy. Next consider the regress of Strong Block obligations:

- I should refrain from buying the shoes;
- I should learn the above obligation;
- I should learn the above obligation;
etc.

This seems harder. For how are we to learn an infinity of such obligations? Nevertheless, this should pose no problem given that all obligations (i.e. after the first) are of the same, simple sort. Furthermore, if I learn the general obligation that I should learn all the above obligations, then I learn at once all obligations described in the regress. Compare: if I learn the general obligation that I should buy all the shoes, then I learn many particular obligations at once, that is, as many obligations as there are shoes.

Importantly, this kind of consideration does not immediately generalize. For example, if I learn the utilitarian principle that I ought always to maximize utility, then, in a certain sense, I have thereby learned many obligations, namely the obligation to maximize utility at t_1, to maximize utility at t_2, and so on. This time, though, I have not learned all I need to learn in order to carry out these obligations. The problem is that I cannot 'just' maximize utility: in each case, I also have to find out what specific actions do maximize utility. Given their simple and unsophisticated content, no such worry applies to the above regress of Strong Block obligations: all there is to carrying out a certain obligation in the list is learning that I have the previous obligation in the list.

Finally consider the regress of possibilities of knowing (in terms of both Block versions):

- I can know I should refrain from buying the shoes;
- I can know I should refrain from making the above obligation unknowable/learn it;
- I can know I should refrain from making the above obligation unknowable/learn it;

etc.

This is not absurd either. The reason is similar to the reason provided above: namely that all propositions I should be able to know (i.e. after the first) are of the very same, straightforward sort. Furthermore, if I am in a position to know the general proposition 'I should refrain from making all of the above obligations unknowable/learn them', then I am in a position to know all of its instances (including those described in the regress). Compare: if I am in a position to know the proposition 'I should buy all the shoes', then I am in the position to know many further propositions at once; as many propositions as there are shoes to buy.

Nevertheless, one might worry that from a certain point in the regress onwards, it is no longer possible to think about the given obligations. For example, is it really possible to think about obligation no. 100,000 to learn obligation no. 99,999? If this is not the case, and assuming that obligations should be comprehensible in order to be knowable, then many of the obligations in the regress will be unknowable after all. In response, it can be pointed out that in order to think about obligation no. 100,000 it is not needed to reason through the whole series until one arrives at it. All that is needed to be able to think about the given obligations, and grasp their content, is to understand the concept of a predecessor.

For example, one merely needs to understand what 'the predecessor of the given obligation' means in order to grasp obligation no. 100,000. But that is easy.

In sum, if none of these consequences are really absurd, then (5) is false, we do not have a paradox, and consequently Access cannot be refuted on the basis of this argument. This concludes my analysis of the IRA spelled out along the lines of the Paradox Theory. Yet there is also another way to make the IRA explicit, a way that involves a different set of issues.

5.4 Failure Analysis

The second theory of IRAs holds not that regresses lead to paradoxes, but that they lead to failures. According to this theory, IRAs demonstrate that solutions to a given problem fail because they get stuck in a regress, namely of similar problems that must be solved in order to solve the initial problem (see Chap. 3). Applied to the Access case, the idea is roughly this: Sider's Block solution fails because it gets stuck in a regress of ever further obligations that have to be secured first, i.e. in order to secure any obligation at all. Let us first consider the full reconstruction of this IRA before explaining its steps[8]:

Access (Failure B instance)

(1) For all obligations x, if I have to demonstrate that the shirker has x, then I refer to her obligation to refrain from making x unknowable.
(2) For all obligations x, if I refer to her obligation y to refrain from making x unknowable, then I first have to demonstrate that she has y in order to demonstrate that she has x.
(3) For all obligations x, if I have to demonstrate that she has x, then I first have to demonstrate that she has the obligation y in order to demonstrate that she has x. [from 1–2]
(4) For all obligations x, I will never demonstrate that she has x. [from 3]
(C) If I refer to a further obligation every time I have to demonstrate that she has one, then I will never demonstrate that the shirker has any obligation. [from 1–4]

As said, the goal of this argument is to demonstrate that Sider's Block fails. The reasoning is as follows: (1) is the Block hypothesis, taken into consideration only in order to derive a failure from it. (1) states that I refer or call the attention to the shirker's obligation to refrain from making x unknowable in order to demonstrate that she has obligation x. Then, the proponent of this argument shows that the proponent of (1) has to concede the premise (2). (3) follows from the foregoing,

---

[8] Note that we are switching from the first-person perspective of the agent to the third-person perspective of an advisor for reasons that will become clear below.

and the failure (4) from (3). Now we have derived (4) from (1), which means that we may conclude 'if (1), then (4)' (by Conditional Proof).

On the basis of the Failure reconstruction it is again clear how the regress could form a problem for Access. This time, the regress is directed against Block, i.e. the proposal to save Access from the Shirker Problem. Furthermore, on the basis of the reconstruction it can easily be seen how Sider could resist the argument. Namely, he could try to deny the premise (2). According to this premise, to continue with the shoes example, I do not demonstrate that the shirker has an obligation to refrain from buying the shoes *unless* I first demonstrate that she has the obligation to refrain from making this obligation unknowable. By parity of reasoning, we could generate an infinite list of such problems that must be solved first:

- I have to demonstrate that the shirker has the obligation to refrain from buying the shoes;
- I *first* have to demonstrate that she has the obligation to refrain from making the first obligation unknowable;
- I *first* have to demonstrate that she has the obligation to refrain from making the second obligation unknowable;

etc.

Generally, if for *any* problem of a certain sort, I first have to solve another problem of the same sort, then I will never solve any such problem. This is the failure referred to in this reconstruction of the argument. Now, is it plausible to deny premise (2), which generates this list? In the following I will first present a scenario in which (2) is plausible, and then discuss three solutions to block (2).

Promise (2). Suppose you are the shirker who wants to buy the shoes, and that in fact you do buy them, and use the loophole in Access to elude your obligation to refrain from buying them. Further suppose that I am the prosecutor or judge who wants to punish you, and finally that obligations are like laws that one has to institute, and that in the initial situation there are just two laws: Access and Law A: Everyone ought to refrain from supporting child labour.

Scenario (1)

I cannot use A against you, as you are a shirker and A is not knowable to you. Yet I still want to punish you and so institute a new law, namely Law B: Everyone ought to refrain from making it impossible to know whether one has A.[9]

Scenario (2)

I cannot use B against you, as you are a shirker and B is not knowable to you. Yet I still want to punish you and so institute a new law, namely Law C: Everyone ought to refrain from making it impossible to know whether one has B.
etc.

---

[9] For this kind of response to work, we assume that the new laws apply retrospectively.

In this set-up (which resembles Sorensen's set-up discussed in Sect. 5.2), premise (2) holds: I have to institute new laws (or secure new obligations) one by one, and thus will never be able to punish you.

Contra (2). Now suppose I am clever, and from the outset institute the general (and self-referential) law:

Weak General Law. S should refrain from making *any* of her obligations unknowable (including this obligation).

This includes all obligations generated in the regress, and so one need not institute new laws all the time: (2) fails.

This strategy is reminiscent of similar ones to short-circuit other regresses. Consider for instance Frege's well-known regress against the correspondence theory of truth (1918, p. 291). According to one reconstruction of the argument, it follows from this theory that one has to solve the following list of problems:

- I have to determine whether the proposition (p_1) that the shoes are made by a child in Indonesia is true;
- I *first* have to determine whether the proposition (p_2) that p_1 corresponds with reality is true;
- I *first* have to determine whether the proposition (p_3) that p_2 corresponds with reality is true;

etc.

From this it follows that I will never determine whether p_1 (or indeed any other proposition) is true. Yet suppose the proponent of the correspondence theory assumes the following:

For all propositions x, x is true iff the proposition that x corresponds with reality is true.

By this equivalence, all propositions in the regress have the same truth conditions, and so their truth can be determined at once. As Dummett states this point (cf. Künne 2003, p. 131; Rescher 2010, pp. 55–6):

[...] there is no harm in this, as long as we recognize that the truth of every statement in this series is determined simultaneously: the regress would be vicious only if it were supposed that, in order to determine the truth of any member of the series, I had first to determine that of the next term in the series (1973, p. 443).

Unfortunately, in the present case, this strategy is too weak. For Weak General Law fails to work against the kind of shirker who *already* has a limited epistemic perspective.[10] Namely, if a shirker never pays any attention to her obligations, she will never be in the position to know that she should not buy the shoes. That is, this obligation will not be knowable to her even if she never made it unknowable. She just never made it accessible to her in the first place. And this suffices to avoid obligations on the basis of Access.

---

[10] See Sect. 5.2 for the distinction between two kinds of shirkers.

To circumvent this problem, we could suggest that one is not obliged to refrain from making one's obligations unknowable, but rather is obliged to learn them:

Strong General Law. S should learn *all* of her obligations (including this obligation).

By this, the argument can again be stopped. Yet, this solution is implausible too. No one has the obligation to study law and philosophy in order to learn all one's obligations, thus shirkers do not have it either. As Sorensen puts the worry:

A genuine solution to this infinite regress might be especially welcomed by ethics instructors. For the curiosity imperative appears to support mandatory ethics courses (1995, p. 255).

Additionally, how can we say that we did *enough* to make our potential obligations knowable? Surely we are not supposed to *optimize* our epistemic perspective, that is, to work ourselves into an ideal, omniscient perspective. For in that case Access would seem altogether idle. Access places an epistemic restriction on our obligations (so that there are no unknowable obligations), and the obligation to learn all obligations says precisely that we should not be restricted in such a way (that all obligations should be knowable).

Now if both solutions to resist premise (2) fail, then it seems the Failure reconstruction is quite a strong argument. Nevertheless, there is a third option we should consider. Namely, we could solve the problem by securing that the shirker has the following obligations:

O1   S should refrain from buying the shoes;
O2   S should refrain from making or keeping O1 or O2 unknowable.

O1 is the base obligation. Moreover, O1 can be replaced with any other base obligation, i.e. an obligation that non-shirkers do have (after all, the problem is to explain that shirkers have the same obligations as non-shirkers). O2 is more complex, and consists in fact of four components:

(i) S should refrain from keeping O1 unknowable;
(ii) S should refrain from making O1 unknowable;
(iii) S should refrain from keeping O2 unknowable;
(iv) S should refrain from making O2 unknowable.

The most substantive component is (i), which implies that S should make O1 (or any other base obligation) knowable. This does not mean that she should actually learn them, but she does have to work herself into such a position that she can easily access them. Component (ii) is straightforward: once S's base obligations are knowable, she is not allowed to make them unknowable again. The final components (iii) and (iv) say that S should make O2 itself knowable, and refrain from making O2 unknowable. In this respect, O2 is self-referential, just like Weak General Law and Strong General Law. The role of these last two clauses is to make sure that shirkers cannot evade their obligations by abusing Access. It is important to emphasize that Access is in force. This means that if O1 or O2 is not knowable to S, then S does not have that obligation. Yet, given the special nature of O2, it can be shown that such a situation will not occur.

There are two cases to be considered. First case: O1 is currently unknowable to S because the relevant information is encrypted (say). By O2's (i), S should make O1 knowable, and cannot evade O1 by keeping O1 unknowable. Clearly, though, this does not suffice, for why would this clause be in force? S could still evade O1 by keeping O2 unknowable (so that O2 is not in force). By O2's (iii) and (iv), however, S is not allowed to keep (or make) O2 unknowable. Hence, S should make O1 knowable and cannot evade this obligation, that is, even if Access is in force. Second case: O1 is currently knowable to S, because the information has been encrypted, and is available in an open access file, yet can be made unknowable by decrypting the information (in such a way that she will no longer be able to access it). By O2's (ii), S is not allowed to do this. Again, this does not immediately suffice, as it has to be shown why this clause is in force. S could still evade O1 by keeping O2 unknowable. Yet, as before, S is not allowed to do this because of O2's (iii) and (iv). Therefore, S cannot evade O1.[11]

Admittedly, just like Strong General Law, O2 does not come for free. Perhaps it does not immediately support mandatory ethics courses (see above), though it does imply that O2 should be knowable to us. This means that we should work ourselves into such a position that we can grasp O2 with its complex content, that is, that we can grasp the obligation to [make and keep our obligations knowable, including this very obligation]. Clearly, however, this problem applies to all complex obligations (given that they should be knowable according to Access, even if they are hard to comprehend), and poses no special worry for my solution.

Before concluding, I would like to make two points. First, I will say something about the nature of this solution. Second, I will address an objection anticipated by Sorensen.

First point. Does my solution block the whole regress, or does it merely stop the regress from being *vicious* (for this distinction, see Sect. 3.2)? This depends, and this is easy to see in Frege's case. On the one hand, one does have to determine the truth of an infinity of propositions, namely the following list:

- The shoes are made by a child in Indonesia;
- The above proposition corresponds with reality;
- The above proposition corresponds with reality;

etc.

On the other hand, with the equivalence at one's disposal, the truth of all these propositions can just be determined by determining the truth of the first proposition. Thus, there is an infinite regress, but it is not vicious. A similar point applies to my solution. On the one hand, one does have to determine that the shirker has an infinity of obligations, namely the following list:

---

[11] Is this solution ad hoc? It seems not: we do not assume O2 for the sole purpose of blocking the regress; rather, we assume it in order to solve the Shirker Problem in a plausible way.

- She should refrain from buying the shoes;
- She should keep the above obligation knowable;
- She should keep the above obligation knowable;

etc.

On the other hand, if one determines that the shirker has O2, then one determines all these obligations at once. Again: if there is an infinite regress, it is non-vicious.[12]

Second point. Sorensen voices a worry against Strong General Law, and this worry may affect O2 as well:

> Can the infinite regress be short-circuited by interpreting the curiosity imperative as self-referential? No, because 'Everyone has a duty to learn his duties—including this very duty' is not self-evident. To infer it from itself would be circular. And to infer it from another proposition reopens the loophole; a shirker can avoid knowledge that the curiosity principle is true by disposing of his opportunities to learn whether it is true (1995, p. 255).

In terms of my solution, the objection is as follows. The solution short-circuits the regress by invoking O2, a self-referential obligation. What is the justification for this obligation? First, O2 is not self-evident and cannot be justified by referring to itself (one cannot justify the obligation to learn O2 by pointing out that O2 tells us to learn O2). Second, this law could be justified by a certain other proposition X. In the latter case, furthermore, it seems that the shirker could evade O2 by making X unknowable. For if X is unknowable, she cannot use X in her justification of O2, and if she cannot justify O2 then she cannot know it, and then according to Access she does not have to comply with it.

In my view, the shirker cannot do this; she cannot evade O2 in this way. I am inclined to think that O2 is justified simply because it is part of a good solution to the Shirker Problem. Consider the law court analogy again: I am the judge, you the shirker; I have instituted O2 and now want to use it against you. However, now suppose you respond that you are not yet in the position to know O2, as I did not yet justify it, i.e. did not yet convince you that it is a good law. Clearly, this response is inappropriate. All I need to secure is that you are in the position to know (and to be justified in believing) *that* you have to comply with O2, not *why* you must do so.

All in all, my proposal is that the IRA discussed in this section can be stopped by Access proponents if they assume that two specific obligations are in place, namely O1 (or any other base obligation) plus O2.

---

[12] In terms of the Failure B reconstruction above, this means that the regress is generated with the help of premise (2), but without the term 'first'.

5.5 Concluding Remarks

In the foregoing, I have made the IRA against the Access principle explicit in two ways: in Paradox format and in Failure format. Next I showed on the basis of the reconstructions how both arguments can be resisted. The Paradox argument fails because the regressive consequences of Access combined with Block are not absurd. The Failure argument fails, as we have just seen, if we assume O1 and O2. If all this is right, then I have substantiated Sider's position in the controversy: the regress of obligations is not harmful for Access. I will conclude with two general remarks about epistemic restrictions, shirkers, and regress problems.

First remark. At certain points, both Sider (1995, p. 279) and Sorensen (1995, p. 255) seem to suggest that we need not take shirkers seriously. For what might it mean, for instance, to violate one's obligation to refrain from making SuperSuperBlock unknowable (which is the obligation that one should not make Super-Block unknowable)?

But I am inclined to disagree here: it is not at all difficult to make SuperSuperBlock unknowable. There are two possible cases: either I am already in the position to know SuperSuperBlock (e.g. I followed elementary philosophy classes and downloaded the articles by Sorensen and Sider), or I am not yet in a position to know SuperSuperBlock (e.g. I have yet to take the elementary classes). In the first case, to make SuperSuperBlock unknowable in the relevant sense is to make sure, first, that I forget about SuperSuperBlock (which is not difficult, given its complex content), and, second, that the articles by Sorensen and Sider are no longer easily accessible to me (e.g. I could remove them from my computer). In the second case, moreover, I merely have to refrain from taking philosophy classes, and that seems easy enough. This shows that it is easy to violate SuperSuperBlock.

Furthermore, if we link our obligations with issues of responsibility, and say that we are responsible (and possibly blameworthy) for what we do only if we are in a position to know our obligations, then, based on the loophole in Access, I could make it the case that I am never responsible and to be blamed for what I do. If Access has this possible consequence, then shirkers should be taken seriously.

Second remark. In my view, some sort of epistemic restriction on our obligations like Access is plausible. Even if one thinks our obligations are not limited in such a way, one might still accept the view that there is an epistemic restriction on the things we are responsible for, and so possibly blameworthy for (see Access for Blame in Sect. 5.1). Yet *all* such restrictions suffer from loophole problems (as the one identified by Sorensen). Namely: if there is an epistemic restriction on X (e.g. obligations, responsibility, blameworthiness, rationality), then so long as we can manipulate our epistemic circumstances, we can manipulate X. This is just what can be expected from Access and kindred principles.

Moreover, if any proposal to block such loopholes is subject to the *very same* epistemic restriction, then regress problems are again only to be expected. The ultimate solution, therefore, is to hold exactly that *something* is immune to epistemic restrictions. In the Access case, as I argued in Sect. 5.4, this 'something' is the following: we have the unrestricted obligation to make and keep our obligations knowable.[13]

References

Driver, J. 2012. What the objective standard is good for. *Oxford Studies in Normative Ethics* 2: 28–44.

Dummett, M. 1973. *Frege: Philosophy of language*. Duckworth: London.

Frege, G. 1918. The thought: A logical inquiry. Trans. A. M. and M. Quinton 1956. *Mind* 65: 289–311.

Graham, P. 2010. In defense of objectivism about moral obligation. *Ethics* 121: 88–115.

Guerrero, A.A. 2007. Don't know, don't kill: Moral ignorance, culpability, and caution. *Philosophical Studies* 136: 59–97.

Künne, W. 2003. *Conceptions of truth*. Oxford: OUP.

Moore, G.E. 1912. *Ethics*. ed. Shaw W.H. 2005. Oxford: OUP.

Rescher, N. 2010. *Infinite regress: The theory and history of a prominent mode of philosophical argumentation*. New Brunswick: Transaction.

Sher, G. 2009. *Who knew? Responsibility without awareness*. Oxford: OUP.

Sider, T. 1995. Sorensen on unknowable obligations. *Utilitas* 7: 273–279.

Smith, H.M. 2010. Subjective rightness. *Social Philosophy & Policy* 27: 64–110.

Sorensen, R.A. 1995. Unknowable obligations. *Utilitas* 7: 247–271.

Wieland, J.W. 2014. Access and the shirker problem. *American Philosophical Quarterly*.

Zimmerman, M.J. 2008. *Living with uncertainty: The moral significance of ignorance*. Cambridge: CUP.

---

[13] This is what I labelled as 'O2'. O2 immune to epistemic restrictions in the sense that it cannot be abused by shirkers. Note that it is not immune to restrictions in the sense that it is subject to itself.